Handbook of Toxic Plants of North America

Handbook of Toxic Plants of North America

George E. Burrows
Ronald J. Tyrl

Blackwell Publishing

George E. Burrows, DVM, PhD, is emeritus professor of toxicology, College of Veterinary Medicine, Oklahoma State University, Stillwater.

Ronald J. Tyrl, PhD, is professor of botany and curator of the herbarium at Oklahoma State University, Stillwater. He is coordinating editor of the Flora of Oklahoma Project.

Blackwell Publishing Professional
2121 State Avenue, Ames, Iowa 50014, USA

Orders: 1-800-862-6657
Office: 1-515-292-0140
Fax: 1-515-292-3348
Web site: www.blackwellprofessional.com

Blackwell Publishing Ltd
9600 Garsington Road, Oxford OX4 2DQ, UK
Tel.: _44 (0)1865 776868

Blackwell Publishing Asia
550 Swanston Street, Carlton, Victoria 3053, Australia
Tel.: +61 (0)3 8359 1011

First edition, 2006

Library of Congress Cataloging-in-Publication Data

Burrows, George E. (George Edward), 1935-
Handbook of toxic plants of North America / George E. Burrows, Ronald J. Tyrl.
p. cm.
ISBN-13: 978-0-8138-0711-9 (alk. paper)
ISBN-10: 0-8138-0711-5 (alk. paper)
1. Veterinary toxicology—North America. 2. Plant toxins—Physiological effect—North America. 3. Poisonous plants—North America. I. Tyrl, Ronald J. II. Title.
SF757.5.B87 2006
636.089'59—dc22

2005022939

The last digit is the print number: 9 8 7 6 5 4 3 2

Contents

Preface

Toxic plants are very much a part of our environment! Although seldom causing problems, the potential for danger is always there. Veterinarians, pathologists, and other individuals who first encounter an affected animal are presented with a suite of clinical and/or pathological signs that they often must relate to a specific toxic plant or plants. This book is written in response to that need. Our objective was to take the information we compiled in *Toxic Plants of North America* (Iowa State University Press, 2001), reorganize it to emphasize disease syndromes, and relate them to the plants responsible. This approach is the opposite of that employed in *Toxic Plants* in which the plants themselves were the primary focus. We hope that this change in focus will make it easier for our readers to learn and recognize the disease syndromes caused by different toxic plants.

Organization and Format

The affected organ or body system serves as the organizational unit in this book. Each chapter is devoted to one organ or system and a synopsis of the toxic plants that affect it. To facilitate access and review, the information is divided into three sections: (1) principal signs of the disease problem, (2) a table of the plants causing the problem, and (3) synopses of the plants and specific aspects of the problems they cause. In each chapter, we group the plants into categories based on the principal signs presented: for example, diarrhea, weight loss, and rumen impaction in chapter 3.

Toxic plants may affect only a single organ or system and produce a well-defined suite of clinical signs and/or pathologic signs, or they may affect several different organs or systems to varying degrees and produce a plethora of clinical signs. Those that affect multiple organs/systems thus appear repeatedly in the initial lists of causative plants in different chapters. To distinguish among the varying degrees of effect, we have classified the causative plants in this book as primary, secondary, or tertiary.

Primary Plants

Primary plants are those that have well-documented effects on a particular organ/system. They produce well-defined suites of clinical and/or pathologic signs that often have descriptive disease names (e.g., "slobbers," "cracker-heel disease," and "trembles"). They frequently cause devastating economic losses. Species that are primary causative plants for two organs/systems are treated in each of the respective chapters.

Secondary Plants
Secondary plants are those that may affect an organ/system, but their effects are less severe or less commonly produced than those listed as primary. They are described in the chapters (as indicated in parentheses) in which they are primary causes.

Tertiary Plants
Tertiary plants are those that have the potential to affect the organ/system, but rarely do so, or do not produce significant effects, or have only caused problems elsewhere in the world. They are not discussed further in this book and the reader is directed to the information about them presented in *Toxic Plants of North America* or other similar texts.

The synopsis of each toxic plant typically comprises summaries of specific clinical signs, diagnostic aids, pathology, treatment, prevention, disease problems and causes, taxonomy, morphology, and ecology. Photographs, line drawings, and distribution maps accompany the textual information.

Use of This Book

We envision the users of this book to be veterinarians, pathologists, veterinary students, animal scientists, agricultural extension agents, farmers, and ranchers; that is, individuals who are likely to be the first to encounter a dead animal or one showing signs of intoxication. We hope that the user, upon encountering an affected animal in the field, will observe carefully the clinical signs it exhibits, review the book's chapter titles, peruse the principal signs, scan the table of plant taxa causing those signs, and survey the pasture for the presence of suspect plants. It is anticipated that if assistance in identifying the plants is needed, agricultural extension agents or taxonomists at nearby universities will be contacted.

Users of this book will undoubtedly note the lack of literature citations and the brevity of the plant synopses. In order to conserve space and to emphasize the disease-syndrome–plant relationship, we have deliberately omitted citations and made the summaries as concise as possible. We envision that our readers wanting or needing additional information will turn to *Toxic Plants of North America*, a book intended to be encyclopedic in scope.

We must offer an obvious caveat regarding use of this book: not all toxic plants are covered. As noted above, we have deliberately selected the plant species most commonly implicated in animal intoxications. Again, our reader is referred to *Toxic Plants* for information about these other plants.

Acknowledgements

As we have written this book, a number of individuals have volunteered encouraging words, their time, and their expertise. It is most appropriate to acknowledge formally their contributions at this point. Special thanks are due the following individuals who kindly granted us use of their striking photographs: Kelly Allred, Louis B. Anella, Terrence E. Bidwell, Wayne Boucher, Paul Buck, Len Coop, Mark Dimmitt, Wayne J. Elisens, Patricia Folley, Bruce W. Hoagland, Ronald E. Masters, Sandra E. Morgan, A. Joseph Pollard, James L. Reveal, Al Schneider, Charles M. Taliaferro,

John N. Trager, Dennis W. Woodland, and Kazuo Yamasaki. Thanks are also due the Arizona-Sonora Desert Museum, cambridge2000, the Oregon State University Jed Colquhoun Photo Collection, the Samuel Roberts Noble Foundation, and the Texas Agricultural Experiment Station at Uvalde for permission to use their photographs.

Handbook of Toxic Plants of North America

1
Plants Affecting the Blood

Although changes in the blood are often clinical signs characteristic of different disease syndromes produced by a variety of toxic plants, only five genera in five quite different families affect the blood directly. In addition, plants that are nitrate accumulators or cyanogenic also cause conspicuous diagnostic changes in the blood, and thus are included in this chapter.

Principal Signs

I. **Hemorrhage:** intractable bleeding from nose, mouth, rectum, or surgical sites; or localized in hematomas
II. **Hemorrhage:** blood from nose and/or bloody diarrhea; decreased platelets and neutrophils
III. **Hemolysis:** urine dark red to brown, lethargy, anemia; Heinz bodies may be present
IV. **Blood Dark Color:** ruminants especially cattle; weakness, labored respiration, dark mucous membranes, collapse; due to nitrate accumulation
V. **Blood Bright Red:** ruminants; weakness, labored respiration, seizures, collapse; rapidly fatal; due to cyanide

Plants Causing Problems

I. **Hemorrhage:** intractable bleeding from nose, mouth, rectum, or surgical sites; or localized in hematomas

Primary	
Melilotus	sweet clover
Tertiary	
Anthoxanthum	vernalgrass
Cheilanthes	lip fern

II. **Hemorrhage:** blood from nose and/or bloody diarrhea; decreased platelets and neutrophils

Primary

Pteridium	bracken fern

III. Hemolysis: urine dark red to brown, lethargy, anemia; Heinz bodies may be present

Primary

Acer	maple
Allium	onion
Brassica	mustard, turnip

IV. Blood Dark Color: ruminants especially cattle; weakness, labored respiration, dark mucous membranes, collapse; due to nitrate accumulation

Primary	
Pennisetum	millet
Sorghum	sorghum
Secondary	
Amaranthus	pigweed (page 270)
Brassica	mustard, turnip (page 14)
Chenopodium	goose-foot (page 276)
Kochia	kochia (page 86)
Tribulus	goathead (page 262)
Tertiary	
Abutilon	velvet leaf
Avena	oats
Beta	beet
Echinochloa	barnyard grass
Hordeum	barley
Lolium	ryegrass
Poaceae	grasses
Salsola	Russian thistle
Salvia	sage
Triticum	wheat
Viguiera	goldeneyes
Zea	maize

V. Blood Bright Red: ruminants; weakness, labored respiration, seizures, collapse; rapidly fatal; due to cyanide

Secondary	
Prunus	cherry (page 188)
Sorghum	sorghum (page 190)
Suckleya	poison suckleya (page 192)
Triglochin	arrowgrass (page 194)
Tertiary	
Acacia	acacia

Amelanchier	serviceberry
Cercocarpus	mountain mahogany
Cynodon	stargrass
Glyceria	mannagrass
Heteromeles	Christmas berry
Sambucus	elderberry
Sorghastrum	Indiangrass
Stillingia	stillingia

I. Hemorrhage: intractable bleeding from nose, mouth, rectum, or surgical sites, or localized in large hematomas

Melilotus
sweet clover
Fabaceae, pea family

Signs and Diagnostic Aids
- cattle, rarely in horses or sheep
- hemorrhages may occur in and affect various body systems including reproductive or nervous systems, or animal may be found dead
- bleeding from surgical sites or accumulation of blood in subcutaneous tissues
- prothrombin time prolonged from 15 to 100–200 seconds, may persist for a week or more after cessation of eating the plant
- plant stem dicoumarol, >20–30 ppm often 50 ppm

Pathology
- large and/or scattered sites of hemorrhage

Treatment
- severe cases: transfusion with 20 ml/kg blood or 10 ml/kg plasma
- vitamin K_1, 1–5 mg/kg b.w. i.v., i.m., or s.c.; daily for a week or more
- vitamin K_3 at double the dose for vitamin K_1
- feed good quality alfalfa
- do not feed pregnant cows hay or silage with *Melilotus* unless it has been evaluated for dicoumarol

Disease Problem
- "sweet clover disease"
- mainly with hay, slight risk with silage
- typically 10–15% morbidity; 50% case fatality rates
- conversion of coumarin, by fungi within partially cured stems, to toxic dicoumarol (bishydroxycoumarin)
- dicoumarol inhibits epoxidase, preventing regeneration of active form of vitamin K_1; depletion of prothrombin and factors VII, IX, X

Taxonomy and Ecology
- two introduced species of toxicologic concern:

M. alba	white sweet clover
M. officinalis	yellow sweet clover

- annual or biennial herbs, stems erect, 1–2 m tall
- vanilla-like scent when dry
- leaves 1-pinnately compound, leaflets 3, margins dentate
- flowers small, pealike, white or yellow
- cultivated and naturalized in waste areas across continent

Melilotus officinalis

M. officinalis

M. officinalis

M. alba

II. Hemorrhage: blood from the nose and/or bloody diarrhea; decreased platelets and neutrophils

Pteridium
bracken fern
Dennstaedtiaceae, bracken family

Signs and Diagnostic Aids
- bloody nasal discharge, bloody feces, decreased appetite, labored respiration, listlessness, increased body temperature
- marked decrease in neutrophils, platelets; increased capillary fragility, bleeding time; defective clot retraction
- enzootic hematuria: intermittent pinkish to brownish or bloody urine, especially last portions of urine stream; in mature cattle, weeks or months between episodes; cystitis with increased protein, white and red blood cells in urine

Pathology
- areas of extensive hemorrhage in various tissues
- enzootic hematuria; bladder with areas of desquamation, foci of hemorrhage, papillomatous growths or neoplasms

Treatment
- provide nursing care; systemic use of antimicrobials
- transfusion of blood for platelets and neutrophils
- enzootic hematuria; administration of antimicrobials for accompanying cystitis

Disease Problem
- bone marrow depression after large amounts (20–25% of diet) of plant eaten for several weeks or more, radiomimetic effect
- severe deficiency of platelets and neutrophils
- ptaquiloside, possibly other illudanes, hydrolyzed to reactive dienones that bind and cleave DNA
- enzootic hematuria results from long-term (months to years) ingestion of small amounts of plant
- also causes thiamine deficiency in horses (page 138)

Taxonomy and Ecology
- one species with four varieties of toxicologic concern:

P. aquilinum	bracken fern, brake fern, bracken, braken, lacy bracken, tailed bracken

- deciduous ferns, forming large, dense colonies
- rhizomes black, woody
- fronds erect, 0.5–1.5 m long, pinnately compound
- sporangia partially concealed by revolute margins
- well-drained soils of open woodlands across continent

Pteridium aquilinum

P. aquilinum

P. aquilinum

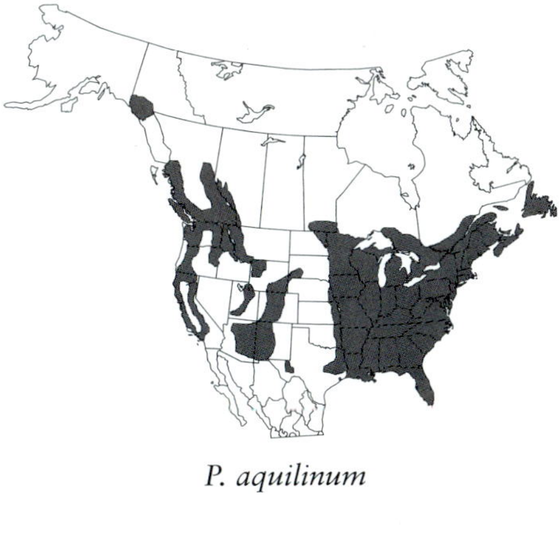

P. aquilinum

III. Hemolysis: horses; urine dark red to brown or none; lethargy, icterus, weakness, anorexia, anemia; Heinz bodies in blood

Acer
maple
Aceraceae, maple family

Signs and Diagnostic Aids
- onset quickly progressive over several days following exposure to broken tree limbs and wilted leaves
- depression, lethargy, anorexia, urine dark red to brown
- hematologic changes: anemia, numerous Heinz bodies in RBCs, MetHb levels up to 50%
- anemia rapid onset or progressive; PCV 8–10%, Hb 5 g/dl
- oliguria and icterus may be primary presenting signs; with severe anemia and indications of renal failure

Pathology
- spleen enlarged, congested with blood
- kidneys swollen, dark reddish-black
- microscopically: tubular nephrosis, Hb casts and cellular debris in dilated tubules

Treatment
- provide supportive care, avoid stress
- administer fluids, whole blood (hemolysis may continue)
- activated charcoal may be of benefit in early stages
- for MetHb, ascorbic acid, 30 mg/kg b.w. i.v. or 3–4 times that amount orally every 12 hours

Disease Problem
- abrupt lysis of RBCs confirmed only in horses
- possibly caused by other maples and may occur in other animal species
- disease appears 1–2 days after eating wilted or dried leaves, 1–3 g/kg b.w.
- fresh green leaves directly from the tree not similarly toxic
- toxicants are associated with formation of reactive oxidants, overwhelm the glutathione-reducing system in RBCs
- intra- and extravascular hemolysis; death due to lack of oxygen delivery to vital cells

Taxonomy and Ecology
- one species of definite toxicologic concern:

A. rubrum	red maple, scarlet maple, swamp maple

- medium-sized tree; bark smooth, gray
- leaves palmately 3–6 lobed, margins coarsely serrate, petioles red
- fruits samaras; bright red, winged
- eastern half of continent; elsewhere as ornamental

Acer rubrum

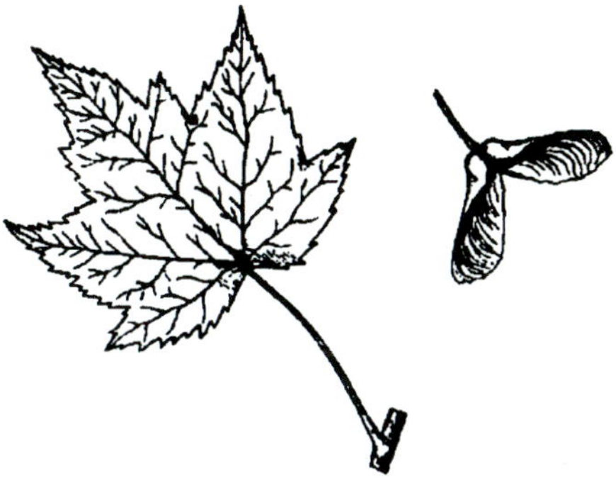

A. rubrum

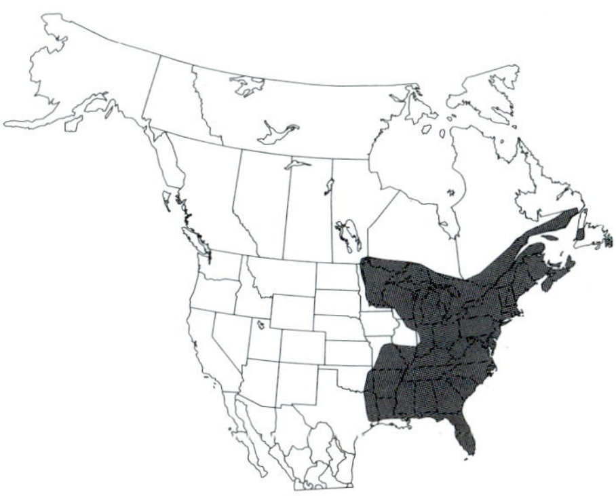

A. rubrum

III. Hemolysis: urine dark red or brown, lethargy, weakness, anorexia, anemia, Heinz bodies, onion odor

Allium
onion
Liliaceae, lily family

Signs and Diagnostic Aids

- abrupt onset; urine dark or coffee colored, lethargy, decreased appetite, weakness
- mucous membranes similar to but darker than with icterus
- strong onion odor from breath or body
- anemia variable but with numerous Heinz bodies in RBCs
- regenerative response: polychromasia, reticulocytes, nucleated RBCs, increased cell size and hemoglobin content, increased neutrophils with numerous bands
- recovery usually prompt and uneventful

Pathology

- liver pale tan to golden brown, perhaps mottled
- kidneys brown to mottled black
- microscopically: mild to moderate hepatic necrosis, renal tubular nephrosis, hemosiderin deposits in liver, kidney, and spleen

Treatment

- typically not required
- blood transfusion rarely needed

Disease Problem

- most toxic to cats and dogs, less in horses, least in cattle and sheep; effects cumulative
- dogs and cats at risk from raw, cooked, or dehydrated onions in soufflés, soups, casseroles, table scraps, or baby food fed to sick cats
- livestock at risk from large amounts of cultivated onions, fresh, spoiled, or cooked, when used as feed or from wild plants
- highly reactive oxidants, from sulfur compounds such as disulfides and thiosulfides (including n-dipropyl disulfide); denaturation of Hb, Heinz body formation, deformation of RBCs, hemolysis, and increased phagocytosis
- low levels of MetHb
- also affects taste of milk and odor of urine

Taxonomy and Ecology

- about 95 cultivated and wild species of toxicologic concern:

  Allium onions, leeks, chives, garlic

- strongly aromatic herbs from bulbs
- leaves basal, long, sheathing
- flowers in flat-topped to globose umbels
- fruits 3-lobed capsules
- variety of habitats; across continent

Allium perdulce

A. perdulce

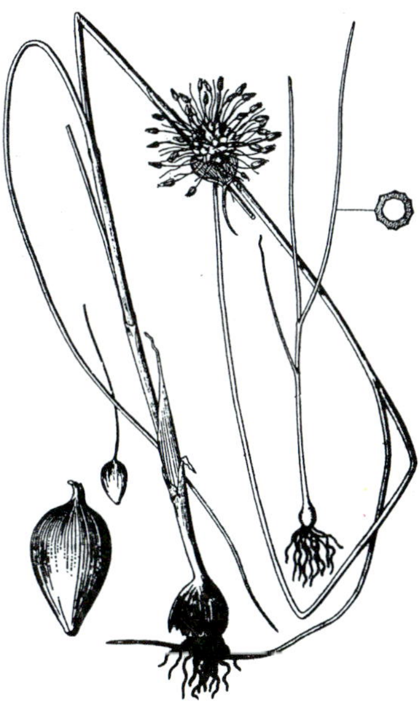

A. cernuum

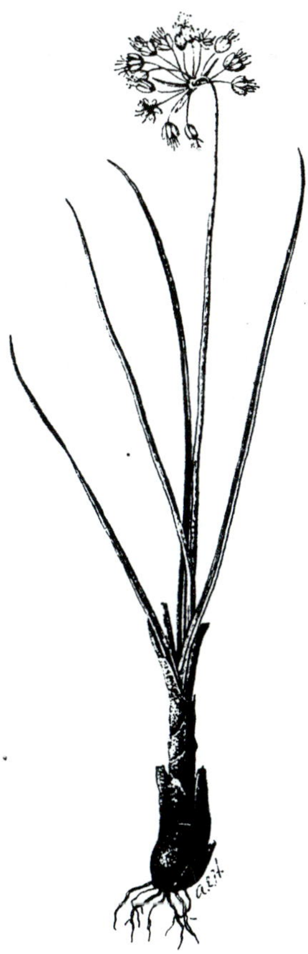

A. vineale

III. Hemolysis: urine dark red to brown, slight to moderate decline in PCV

Brassica
mustard, turnip
Brassicaceae, mustard family

Signs and Diagnostic Aids
- few, if any, distinct signs
- urine dark, PCV decline up to 30% in cattle, less in sheep
- temporary; mild loss of appetite, weakness

Pathology
- seldom fatal but spleen enlarged and kidneys dark metallic in color
- iron and hemosiderin in liver and kidney

Treatment
- seldom needed

Disease Problem
- anemia after eating cultivated species for 2–3 weeks
- sulfides, disulfides, trisulfides, thiosulfonates from S-methyl-L-cysteine sulfoxide (SMCO); adverse effects mainly due to dimethylsulfide or its oxide
- formation of toxicants does not occur in acidic environment of monogastric stomach
- reaction products react with SH groups of Hb; lysis of RBCs directly or secondary to denaturing; Heinz bodies present
- rarely, temporary blindness of unknown cause and not associated with other severe signs may occur in cattle
- also causes ARDS (page 236), PEM (page 117), primary photosensitization (page 244)

Taxonomy and Ecology
- cultivated species of primary toxicologic concern:

B. napus	rape
B. oleracea	kale, broccoli, cauliflower, cabbage, brussels sprouts
B. rapa	turnip, especially tops

- annual or perennial herbs; stems erect, 1–15 cm tall, branched, smooth
- leaves simple, alternate, basal, and cauline
- flowers borne in racemes, conspicuous; petals 4, yellow or yellow-white, free
- stamens 6: 4 long and 2 short

Brassica oleracea

B. nigra

B. nigra

IV. Blood Dark Color: weakness, labored respiration, collapse, dark mucous membranes; ruminants, especially cattle; due to nitrate accumulation

Pennisetum
millet
Poaceae, grass family

Signs and Diagnostic Aids
- abrupt onset; minutes or hours after eating hay
- may be fatal in minutes or a few hours
- weakness, labored respiration, possibly dark sclera and mucous membranes
- plant or animal tissues: test for excess nitrate, drop of 0.5% diphenylamine solution to cut stem, or dilute to 0.1%, several drops to drop of ocular fluid, dark blue color develops quickly

Pathology
- few distinctive lesions
- blood and tissues may be dark or chocolate-colored

Treatment
- administer methylene blue, 2–4 mg/kg b.w. as a 1–2% solution
- use feeding strategies to dilute amount of toxic material or allow animal selection of less-toxic leafy portions
- following prolonged dry period, delay harvest of forage for two weeks after first substantial rain
- adapt animal using oral inoculum paste of *Propionibacterium* such as Bova-Pro®

Disease Problem
- due to nitrate accumulation in stems related to environmental conditions
- associated mainly with hay, fresh forage is of considerably less risk
- conversion of nitrate to nitrite in rumen; absorption of nitrite resulting in oxidation of hemoglobin to MetHb, which is incapable of oxygen transport

Taxonomy and Ecology
- one cultivated species of toxicologic concern:

  P. glaucum pearl millet, horse millet

- robust annual, 1–2.5 m tall
- leaves numerous, wide, long
- inflorescences spicate panicles of spikelets; bristles as long as spikelets

Pennisetum glaucum

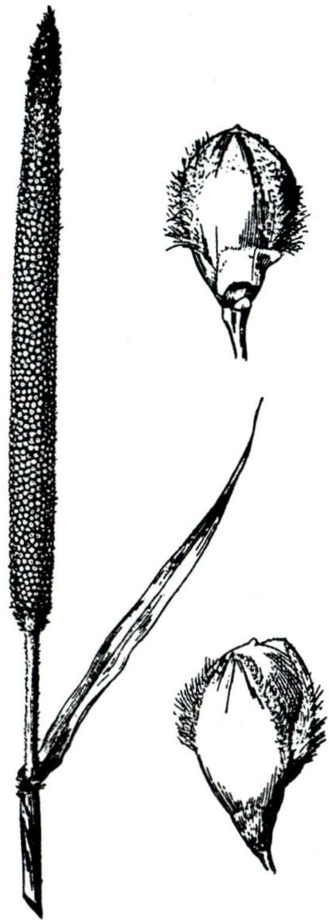

P. glaucum

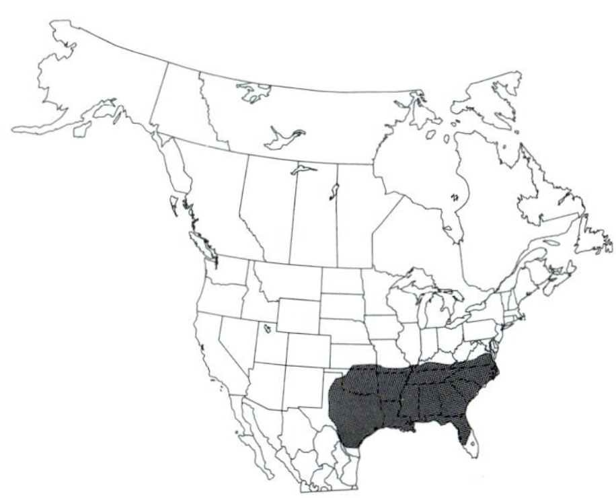

P. glaucum

IV. Blood Dark Color: weakness, labored respiration, collapse, dark mucous membranes; ruminants especially cattle; due to nitrate accumulation

Sorghum
sorghum, sudan
Poaceae, grass family

Signs and Diagnostic Aids
- abrupt onset; minutes or hours after eating hay
- may be fatal in minutes or a few hours
- weakness, labored respiration, possibly dark sclera and mucous membranes
- plant or animal tissues: test for excess nitrate, drop of 0.5% diphenylamine solution to cut stem, or dilute to 0.1%, several drops to drop of ocular fluid, dark blue color develops quickly

Pathology
- few distinctive lesions
- blood and tissues may be dark or chocolate-colored

Treatment and Prevention
- methylene blue, 2–4 mg/kg b.w. as a 1–2% solution
- use feeding strategies to dilute amount of toxic material or allow animal selection of less toxic leafy portions
- following prolonged dry period, delay harvest of forage for two weeks after first substantial rain
- adapt animal using oral inoculum paste of *Propionibacterium* such as Bova-Pro®

Disease Problem
- due to nitrate accumulation in stems related to environmental conditions
- associated mainly with hay, fresh forage is of considerably less risk
- conversion of nitrate to nitrite in rumen; absorption of nitrite resulting in oxidation of hemoglobin to MetHb; incapable of oxygen transport
- also occasionally causes ataxia, cystitis, and urinary incontinence in horses grazing or eating fresh cut hay for several weeks or months

Taxonomy and Ecology
- three species of toxicologic concern:

S. x *almum*	sorghum almum, Columbus grass
S. bicolor	sorghum, shattercane, sudan, milo, sudangrass
S. halepense	Johnsongrass

- cultivated annuals or perennials
- leaves cauline, flat, 12–120 mm wide
- inflorescences; panicles of paired spikelets, one pedicelled and one sesile
- spikelets dorsally compressed, glumes equal
- perennial *S. halepense* is both cultivated and naturalized across the southern half of continent

Sorghum halepense

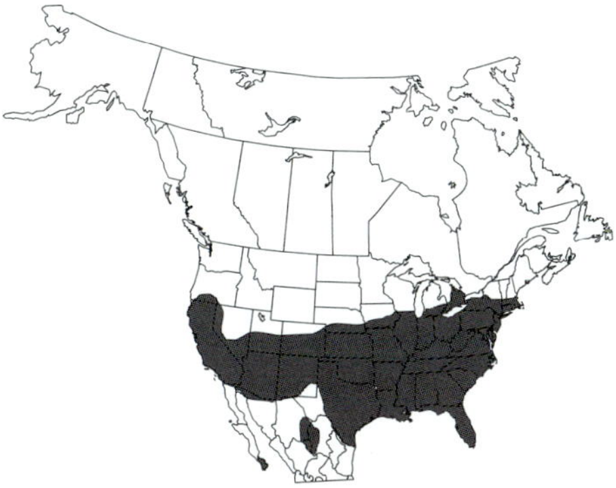

S. halepense

S. halepense

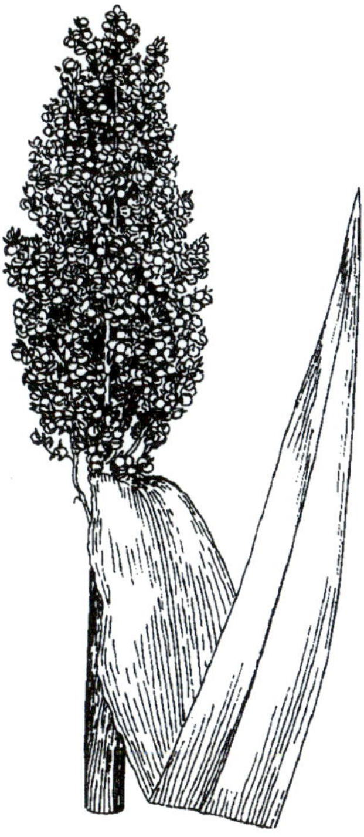

S. bicolor

2
Plants Affecting the Cardiovascular System

As might be expected, the cardiovascular system, and especially the heart, is affected by an extensive array of toxic compounds produced by a variety of plants. Thirty-six genera in 17 families affect the system to varying degrees. Eight of these genera are of significant concern because they are capable of producing serious disease problems and devastating losses. The three categories of principal signs presented reflect the likelihood and rapidity of cardiac failure or peripheral vascular effects and the chemical and physical changes occurring in the animal's body.

Principal Signs

I. **Cardiac Failure Due to Physiologic Effects:** heart rate decreased and/or irregular, weakness, depression, and/or sudden death
II. **Cardiac Failure Due to Cardiomyopathy:** edema, often sudden death, possibly with irregular heart rate
III. **Laminitis and/or Limb Edema**

Plants Causing Problems

I. **Cardiac Failure Due to Physiologic Effects:** heart rate decreased and/or irregular, weakness, depression, and/or sudden death

Primary

Asclepias	nonwhorled milkweed
Digitalis	foxglove
Nerium	oleander
Taxus	yew

Secondary

Kalanchoe	kalanchoe (page 48)
Kalmia	laurel (page 78)
Rhododendron	rhododendron, azalea (page 80)
Sophora	mescal bean (page 184)

Veratrum	false hellebore (page 232)
Tertiary	
Aconitum	monkshood
Andromeda	andromeda
Apocynum	dogbane
Convallaria	lily-of-the-valley
Helleborus	hellebore
Ledum	Labrador tea
Leucothoe	leucothoe
Lyonia	maleberry
Menziesia	menziesia
Pieris	lily-of-the-valley bush
Robinia	black locust

II. Cardiac Failure Due to Cardiomyopathy: edema, often sudden death, possibly with irregular heart rate

Primary	
Gossypium	cotton; cottonseed meal
Secondary	
Astragalus	locoweed (page 128)
Eupatorium	white snakeroot (page 206)
Oxytropis	locoweed (page 136)
Persea	avocado (page 218)
Phalaris	canarygrass (page 202)
Senna	senna (page 60)
Tertiary	
Amaranthus	pigweed
Selenium plants	

III. Laminitis and/or Limb Edema

Primary	
Berteroa	hoary alyssum
Juglans	black walnut
Secondary	
Festuca	fescue, suspected (page 84, 216)
Tertiary	
Lesquerella	bladderpod

I. Cardiac Failure Due to Physiologic Effects: heart rate decreased and/or irregular, weakness, depression, respiration labored, diarrhea, and/or sudden death

Asclepias
opposite-leaved milkweeds
Asclepiadaceae, milkweed family

Signs
- depression, weakness, lying down, reluctance to stand
- abdominal pain, odontoprisis, diarrhea
- irregular labored respiration with a groaning grunt
- intermittent muscular tremors
- arrhythmias typical of digitalis glycosides, seldom seen

Pathology
- reddening of mucosa of stomach and small intestines
- microscopically: mild myocardial cellular degeneration

Treatment and Control
- administer activated charcoal orally
- administer atropine for AV block and/or antiarrhythmic drugs

Disease Problem
- opposite-leaved species are cardiotoxic; whorled-leaved species are neurotoxic
- most animal species are at risk
- fresh plants seldom eaten but remain toxic in hay
- 1–2% b.w. may cause intoxication
- risk highest in early growth; lowest in late season or after first frosts
- 23-C steroidal glycosidic cardenolides
- inhibition of Na^+/K^+ ATPase, movement of Na^+, Ca^{++} into cells
- also affects digestive and respiratory systems

Taxonomy and Ecology
- representative species of toxicologic concern:

A. asperula	antelopehorn
A. brachystephana	short-crown milkweed
A. eriocarpa	woolly-pod milkweed
A. labriformis	labriform milkweed
A. latifolia	broad-leaf milkweed
A. speciosa	showy milkweed
A. syriaca	common milkweed
A. viridis	green milkweed, spider milkweed

- herbs with white, latexlike sap, 4–150 cm tall
- leaves opposite, typically wide
- flowers in umbellate clusters
- petals of various colors
- coronas conspicuous, hood-shaped
- fruits globose, fusiform or ovoid follicles
- seeds flat, with tuft of hairs
- variety of habitats; most vegetation types; greatest abundance southward

Asclepias speciosa

A. latifolia

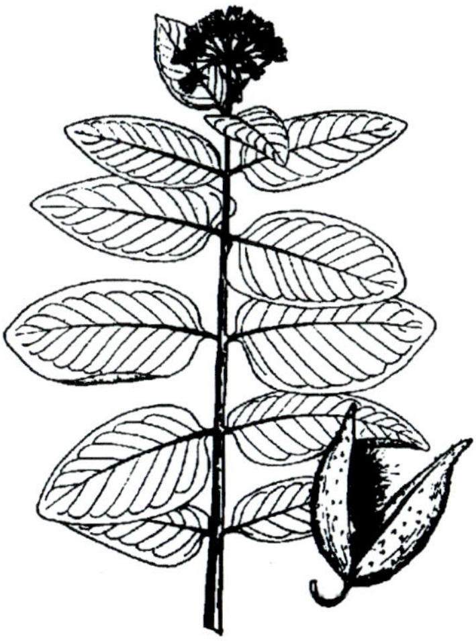

A. speciosa

A. syriaca

I. Cardiac Failure Due to Physiologic Effects: heart rate decreased and/or irregular, weakness, diarrhea, respiration labored, distinct arrhythmias

Digitalis
foxglove
Scrophulariaceae, snapdragon family

Signs and Diagnostic Aids

- earliest are marked weakness, stupor, diarrhea (± blood)
- sheep, cattle, and deer: colic, labored respiration with a peculiar forced expiratory sound
- heartbeat loud, strong; later slow, irregular, almost imperceptible; eventually ventricular fibrillation
- increased frequency of urination
- dogs and cats: vomiting, salivation, little diarrhea
- diagnosis facilitated by measurement of ATPase activity in RBCs or by radioimmunoassay for digitoxin in serum

Pathology

- reddening of mucosa of stomach and intestines
- heart dilated
- microscopically: swollen, fragmented cardiac muscle cells
- presence of leaf fragments in stomach or intestine

Treatment

- administer activated charcoal orally
- administer atropine, with added propranalol as needed
- use digoxin-specific Fab fragment antibodies

Disease Problem

- numerous 23-C steroidal glycosidic cardenolides
- inhibit Na^+, K^+ ATPase, increase intracellular calcium, cause conduction problems, AV block, ventricular fibrillation
- water from vases with fresh flowers is a risk
- also affects digestive and urinary systems

Taxonomy and Ecology

- three species of toxicologic concern:

D. purpurea	purple foxglove, common foxglove
D. lanata	Grecian foxglove
D. lutea	straw foxglove

- erect rarely branched herbs, 30–180 cm tall
- leaves alternate with lower crowded on stems
- flowers in elongate terminal racemes, showy, tubular
- petals purple, pink, or white, throat spotted
- *D. purpurea* common ornamental; naturalized in Pacific Northwest, moist areas, open pastures
- *D. lanata* and *D. lutea* rare in Northeast

Digitalis purpurea

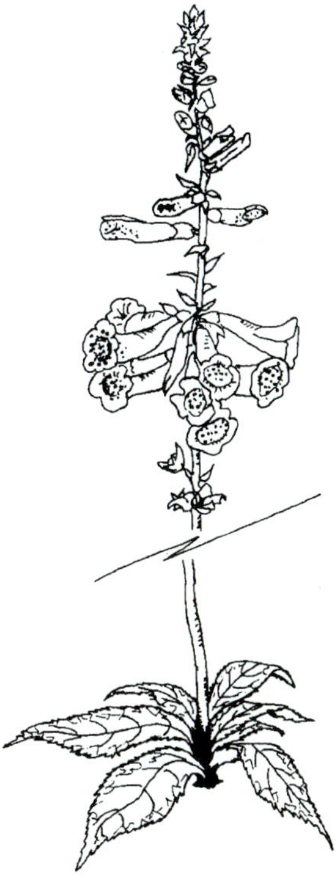

D. purpurea

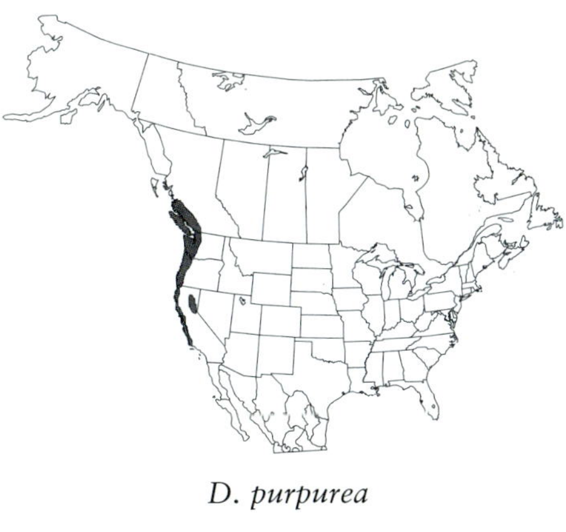

D. purpurea

I. Cardiac Failure Due to Physiologic Effects: heart rate decreased and/or irregular, weakness, depression, salivation, distinct arrhythmias

Nerium
oleander
Apocynaceae, dogbane family

Signs and Diagnostic Aids
- weakness, depression, anorexia, excess salivation, abdominal pain
- diarrhea (± blood) consistent in all animals
- vomiting common in dogs and cats
- cardiac signs: bradycardia, paroxysmal tachycardia, S-T segment depression, arrhythmias such as AV block, ectopic beats, gallop rhythm with dropped beats
- onset of signs delayed up to 12 hours in cattle and sheep
- flaring, soreness around mouth, colic-type signs in horses
- if animals are not periodically and carefully observed may exhibit sudden death with few premonitory signs
- occasionally marked hyperkalemia

Pathology
- reddening of mucosa of stomach and intestines
- tough, leathery leaves persist in stomach; uniformly spaced secondary veins resemble ribs of trout skeleton, indistinct near curled margins
- microscopically: myocardial necrosis and fibrosis
- diagnosis aided by analyses of gut contents, urine

Treatment
- administer activated charcoal orally
- administer atropine and propranolol as needed
- use digoxin-specific Fab antibody fragments

Disease Problem
- seldom eaten when fresh, dried leaves main risk
- most animal species, including birds, wildlife, and zoo animals are at risk
- less than 0.5% body weight may cause intoxication
- 23-C steroidal glycosidic cardenolides
- inhibit Na^+, K^+ ATPase, increase intracellular calcium, cause conduction problems, AV block, ventricular fibrillation
- water from vases with fresh flowers is risk to pets
- also affects digestive system

Taxonomy and Ecology
- one polymorphic species of toxicologic concern:

N. oleander	oleander, laurel rosa, laurel blanco

- evergreen shrub, 2–6 m tall
- leaves simple, whorled, or opposite, leathery
- flowers fragrant, funnelform
- petals 5-lobed, white, pink, or purple
- fruits long follicles, numerous feathery seeds
- widely planted ornamental in southern states and Mexico; numerous cultivars; naturalized in some areas

Nerium oleander

N. oleander

I. Cardiac Failure Due to Physiologic Effects: few premonitory signs, trembling, collapse, heart rate decreased and/or irregular or sudden death

Taxus
yew
Taxaceae, yew family

Signs
- most cases: sudden death, especially in horses, preceded only by brief period of several minutes of trembling and quivering
- cattle show vague signs, most commonly found dead
- death few minutes after eating plant or days later

Pathology
- animals usually die too quickly for development of lesions
- presence of leaves and leaf fragments in rumen or stomach
- analysis of ingesta for taxine or related compounds

Treatment
- administer atropine when ingestion observed but signs not yet apparent
- lidocaine use is questionable because it requires monitoring, and repeated i.v. administration
- activated charcoal orally may be used as a preventive

Disease Problem
- extremely hazardous plant, all parts toxic except fleshy aril of fruit
- both fresh and dried leaves dangerous; less than 0.1% b.w. may cause intoxication
- most animal species including dogs, cats, other house pets are at risk
- taxines A, B; block ion channels; interfere with cardiac conduction causing heart block, slowing of heart, distinctive widening of QRS

Taxonomy and Ecology
- introduced ornamentals of toxicologic concern; native species not a problem:

T. baccata	English yew
T. chinensis	Chinese yew
T. cuspidata	Japanese yew
T. x *media*	Hicks yew

- evergreen trees or shrubs
- leaves needlelike, flexible, flattened, often 2-ranked
- producing pollen and seed cones
- seeds borne in red-orange/scarlet-red, fleshy cup-shaped arils
- widely planted as hedges and ornamentals

Taxus cuspidata

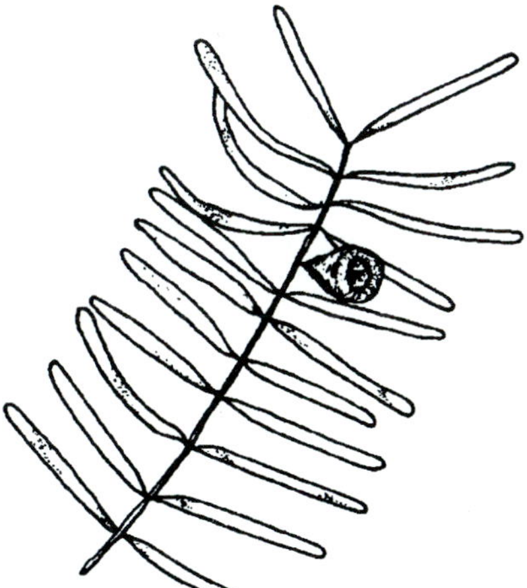

Taxus

II. Cardiac Failure Due to Cardiomyopathy: generalized edema, weakness, frothy salivation, respiratory distress

Gossypium
cottonseed
Malvaceae, mallow family

Signs and Diagnostic Aids

- after consumption of cottonseed meal for weeks or months; abrupt onset of muscular weakness, frothy salivation, generalized edema and hypothermia
- pigs: respiratory distress, gasping/thumping respiration
- signs of cardiac insufficiency, death 3–7 days after onset
- calves and lambs: abrupt onset of vague but severe signs; erratic appetite, muscular weakness, hypothermia, generalized edema, respiratory distress and/or Hburia
- adult cattle: increased RBC fragility, impaired testicular function
- all animals: often death without prominent premonitory signs
- increase of serum CPK, LDH; decrease of serum Ca, Hb, PCV

Pathology

- widespread congestion and edema
- reddening and sloughing of mucosal surface of stomach and intestines
- large amounts of free fluid in chest and abdominal cavities, pulmonary edema
- liver mottled or nutmeg appearance, heart dilated, flabby
- microscopically: interstitial cardiac edema with separation and degeneration of myofibers; diffuse centrilobular necrosis of liver; mild kidney degeneration

Treatment and Prevention

- cumulative nature of toxicant and pathological changes preclude treatment; deaths continue several days after cessation of feeding cottonseed meal
- proper dietary management most effective: free-gossypol <100 ppm for pigs, lambs, calves; average gossypol content of meal limit to <9% of total ration for pigs

Disease Problem

- cottonseed meal is residue from pressing process, used as a protein feed supplement
- cumulative effects; case fatality rate typically >50%
- mainly swine, but also preruminant calves and lambs
- mature ruminants at risk when fed very high levels
- gossypol, highly reactive binaphthalene, in distinctive oil glands in cotyledons of seed; free form toxic
- mainly protein-complexed and inactive in functional rumen
- also reproductive effects; spermatozoa poor and few

Taxonomy and Ecology

- two cultivated species of toxicologic concern:

G. barbadense	sea-island cotton
G. hirsutum	upland cotton

- erect subshrubs or shrubs
- leaves alternate, palmately lobed, petiolate
- flowers large; showy; petals yellow, cream, or mauve
- cultivated; southern half of continent

Gossypium hirsutum

G. hirsutum

III. Laminitis and/or Limb Edema: horses; abrupt onset of limb edema, increased foot temperature

Berteroa
hoary alyssum
Brassicaceae, mustard family

Signs and Diagnostic Aids

- increased foot temperature 24 hours after ingestion
- abrupt onset; pitting edema of distal portions of one or more limbs; in most cases extent of disease
- some animals develop mild to severe laminitis in 1–2 days; stand with forelegs forward and weight on pelvic limbs
- when plant amounts high: colic, bloody diarrhea, slow capillary refill time, fever to 40°C, increased heart and respiration rate, dehydration; urine may be red
- PCV may increase to greater than 50%, with increase in serum creatinine, urea nitrogen, and total protein

Pathology

- subcutaneous edema, pulmonary edema, edema of kidney
- mucosa of stomach and small intestine may be ulcerated, small splotchy hemorrhages on mucosal and serosal surfaces
- large amounts of calcium carbonate crystals in bladder

Treatment

- provide relief of pain, inflammation, and edema: phenylbutazone, flunixin megumine, furosimide, or other diuretics

Disease Problem

- only in horses; upper Midwest
- in heavily infested pastures or hay with 20–30% of plant
- morbidity rate may be higher than 50% but most recover if ingestion ceases immediately
- possible abortion at 9–10 months of gestation
- toxin unknown, but apparently highest concentration in seeds

Taxonomy and Ecology

- one species of toxicologic concern:

B. incana hoary alyssum, hoary false alyssum

- stiffly erect annual herb, 10–70 cm tall
- leaves simple, sessile
- inflorescence raceme of small, white flowers
- flowers petals 4, stamens 6
- fruit elliptic silicles
- seeds round, brown, winged
- disturbed waste areas and meadows; northern portion of continent

Berteroa incana

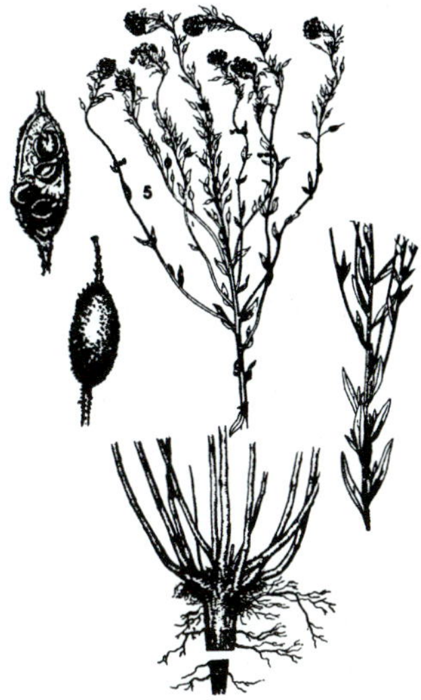

B. incana

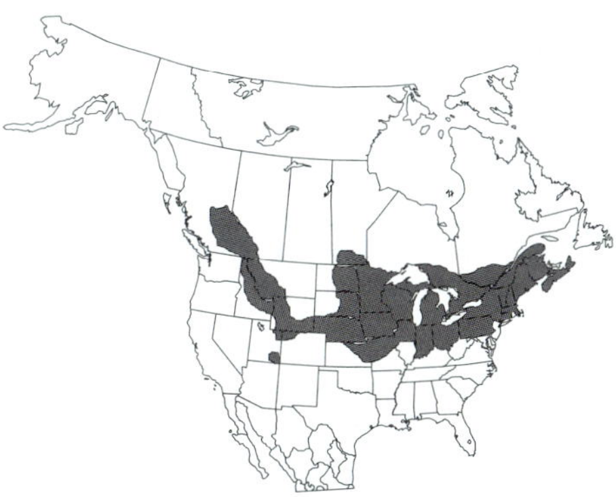

B. incana

III. Laminitis and/or Limb Edema: horses; abrupt onset of depression, limb edema, laminitis, distress

Juglans
black walnut
Juglandaceae, walnut family

Signs and Diagnostic Aids
- horses only: abrupt onset of depression, edema of lower limbs, signs of distress such as flaring of nostrils, colic signs
- laminitis with weight bearing on pelvic limbs, reluctance to move, stiff gait
- ventral edema, increased body temperature, increased respiration and cardiac rates
- with severe laminitis, eventual lameness due to rotation of third phalanx within hoof
- identification of hardwood shavings in bedding

Pathology
- typically a disabling rather than fatal disease

Treatment and Prevention
- immediate removal of the offending bedding-shavings
- administer nonsteroidal anti-inflammatory drugs; phenylbutazone or flunixin meglumine i.v.
- possibly administer adrenergic blockers (prazocin) and/or calcium channel blockers (nifedipine)
- administer mineral oil orally to promote elimination of gut contents and relieve discomfort
- allow shavings to age for weeks prior to use

Disease Problem
- occurs when 20% or more of *Juglans* heartwood contaminates shavings used for bedding
- signs usually seen within one to two days after use of new shavings
- toxin unknown
- possible vasospasm resulting in ischemia within capillaries of laminae with subsequent reperfusion

Taxonomy and Ecology
- one species of toxicologic concern:

  J. nigra black walnut

- large trees
- leaves compound, leaflets 5–25, leaf scars heart-shaped
- fruits nuts enclosed in husks
- bottomlands; eastern half of continent

Juglans nigra

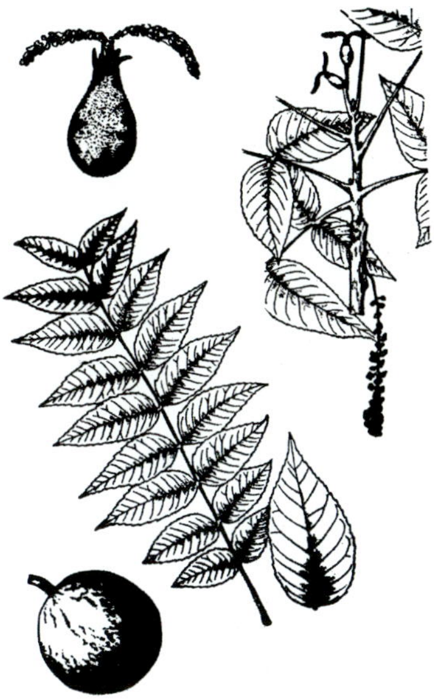

J. nigra

J. nigra

3
Plants Affecting the Digestive System

Frequently, the chemical or physical properties of the compounds produced by toxic plants produce physiologic or pathological signs, such as diarrhea, vomiting, mucosal reddening, or ulcerations, referable to the digestive tract even though their primary effect is on another system of the animal. The appearance of these signs often confirms the primary diagnostic signs exhibited by the affected system. However, 23 genera in 15 families produce toxicants that specifically affect the digestive system. An additional 28 genera produce significance changes in the system in addition to their primary effects on other systems.

Principal Signs

I. **Diarrhea:** often with vomiting and/or salivation
II. **Impaction of Rumen or Stomach**
III. **Rectal Straining or Tenesmus:** often with diarrhea
IV. **Salivation:** greenish saliva and nasal discharge
V. **Salivation:** profuse, with inflammation of mouth and lips
VI. **Salivation:** profuse, without inflammation of mouth and lips
VII. **Salivation:** profuse, with retching/vomiting, colic
VIII. **Salivation:** profuse, with neurologic signs including those from polioencephalomalacia (PEM)
IX. **Weight Loss:** conspicuous, often without other obvious signs

Plants Causing Problems

I. **Diarrhea:** often with vomiting and/or salivation

Primary

Glottidium	bagpod
Iris	iris
Kalanchoe	kalanchoe
Ligustrum	privet

Narcissus	daffodil
Phytolacca	pokeweed
Ranunculus	buttercup, crowfoot
Ricinus	castor bean
Senna	senna, cassia
Sesbania	rattlebush
Secondary	
Artemisia	sagebrush (page 126)
Asclepias	milkweed (page 24, 164)
Baccharis	baccharis (page 168)
Digitalis	foxglove (page 26)
Nerium	oleander (page 28)
Quercus	oak (page 274)
Solanum	nightshade (page 142)
Tertiary	
Abrus	precatory pea
Aconitum	monkshood
Actaea	baneberry
Anagallis	pimpernel
Baptisia	wild indigo
Buxus	box, boxwood
Caltha	marsh marigold
Centaurium	mountain pink
Colchicum	autumn crocus
Cyclamen	cyclamen
Datisca	durango root
Drymaria	drymary
Gladiolus	gladiola
Gymnocladus	coffee tree
Hedera	ivy
Helleborus	hellebore
Hyacinthus	hyacinth
Ilex	holly
Jatropha	nettle spurge
Lobelia	lobelia
Lycopersicon	tomato
Ornithogalum	star-of-Bethlehem
Phoradendron	mistletoe
Podophyllum	mayapple
Rhamnus	buckthorn
Rheum	rhubarb
Robinia	black locust
Sambucus	elderberry
Sapium	tallow tree
Sinapis	mustard
Thlaspi	fanweed
Wisteria	wisteria

II. Impaction of Rumen or Stomach

Primary	
Diospyros	persimmon
Tertiary	
Apocynum	dogbane
Prosopis	mesquite

III. Rectal Straining or Tenesmus: often with diarrhea

Secondary	
Crotalaria	rattlebox (page 106)
Cynoglossum	hound's tongue (page 108)
Phyllanthus	leaf-flower (page 102)
Quercus	oak (page 274)
Senecio	groundsel (page 110)
Senna	senna (page 60)
Tertiary	
Amsinckia	fiddleneck
Buxus	box, boxwood
Colchicum	autumn crocus
Echium	blueweed
Heliotropium	heliotrope

IV. Salivation: greenish saliva and nasal discharge

Primary	
Helenium	sneezeweed
Hymenoxys	bitterweed
Tertiary	
Baileya	desert marigold
Psilostrophe	paperflower

V. Salivation: profuse, with inflammation of mouth and lips

Primary	
Dieffenbachia	dumb cane
Epipremnum	devil's ivy
Philodendron	philodendron
Tertiary	
Alocasia	elephant's ear
Arisaema	jack-in-the-pulpit
Calla	arum
Capsicum	pepper
Daphne	daphne
Dirca	leatherwood
Lysichiton	skunk cabbage

Monstera	Swiss cheese plant
Poaceae	grasses, many species
Podophyllum	mayapple
Sinapis	mustard
Symplocarpus	skunk cabbage
Zantedeschia	calla lily

VI. Salivation: profuse, without inflammation of mouth and lips

Primary	
Trifolium	clover
Secondary	
Prosopis	mesquite (page 156)

VII. Salivation: profuse, with retching/vomiting, colic

Primary	
Kalmia	laurel
Rhododendron	rhododendron, azalea
Zigadenus	death camas
Secondary	
Cycas	sago palm (page 98)
Phytolacca	pokeweed (page 54)
Ricinus	castor bean (page 58)
Urtica	stinging nettle (page 248)
Veratrum	false hellebore (page 232)
Tertiary	
Andromeda	andromeda
Colchicum	autumn crocus
Ledum	Labrador tea
Leucothoe	leucothoe
Lyonia	maleberry
Menziesia	menziezia
Pieris	lily-of-the-valley bush
Tulipa	tulip
Zamia	coontie

VIII. Salivation: profuse, with neurologic signs, including those from polioencephalomalacia (PEM)

Secondary	
Astragalus	locoweed (page 128, 220)
Brassica	mustard, turnip (page 14, 160)
Kochia	kochia (page 86, 160)
Laburnum	golden-chain-tree (page 132)
Lupinus	lupine (page 180, 230)
Melia	chinaberry (page 182)

Tertiary	
Conyza	horseweed

IX. Weight Loss: conspicuous, without other obvious signs

Primary	
Festuca	fescue
Kochia	kochia
Secondary	
Acacia	guajillo (page 122)
Astragalus	locoweed (page 128, 220)
Centaurea	knapweed (page 154)
Crotalaria	rattlebox (page 106)
Cynoglossum	hound's tongue (page 108)
Lolium	perennial ryegrass (page 198)
Oxytropis	locoweed (page 136)
Prosopis	mesquite (page 156)
Pteridium	bracken fern (page 8, 138)
Senecio	groundsel (page 110)
Tertiary	
Amsinckia	fiddleneck
Echium	blueweed, buggloss
Heliotropium	heliotrope
Leucaena	lead tree

I. Diarrhea: profuse, rapid death; usually winter problem

Glottidium
bagpod, bladderpod
Fabaceae, pea family

Signs and Diagnostic Aids
- mainly cattle
- abrupt onset of watery to mucoid diarrhea ± blood, anorexia, ruminal atony, weak, depressed, abdominal pain
- animal stands with back arched, head down, neck extended
- may find animals dead with no premonitory signs observed
- neutrophilia with many immature forms
- chickens: depression, purple comb, muscular twitching, profuse diarrhea

Pathology
- moderate to severe abomasal and intestinal reddening
- renal tubular degeneration, necrosis
- presence of intact or sprouted beans in ruminal contents
- chickens: necrotic enteritis, gizzard and duodenum; fatty degeneration of liver, crop fluid filled, intestines gas filled, feathers pull out easily

Treatment and Prevention
- administer activated charcoal orally
- administer fluids and electrolytes
- feed well before releasing in heavily infested pastures

Disease Problem
- only green and ripe seeds (beans) are toxic, foliage is of little risk
- 1–2% b.w. is toxic depending on how thoroughly the beans are chewed
- mainly animals newly introduced into area with abundant plants or when snow covers normal forage; avid consumption of large amounts of seeds
- typical occurrence in late fall or winter
- after first few days in pasture animals no longer eat seeds
- sesbanimides, especially A; saponins may contribute

Taxonomy and Ecology
- one species of toxicologic concern:

G. vesicarium	bagpod, bladderpod, coffee bean

- erect, annual herb, 1.5–3 m tall
- leaves 1-pinnately compound, lack terminal leaflets
- flowers yellow, pealike, petals 5
- legumes inflated, persistent baglike endocarp around seeds
- often large populations in low sandy areas subject to flooding; southeastern quarter of continent

Glottidium vesicarium

G. vesicarium

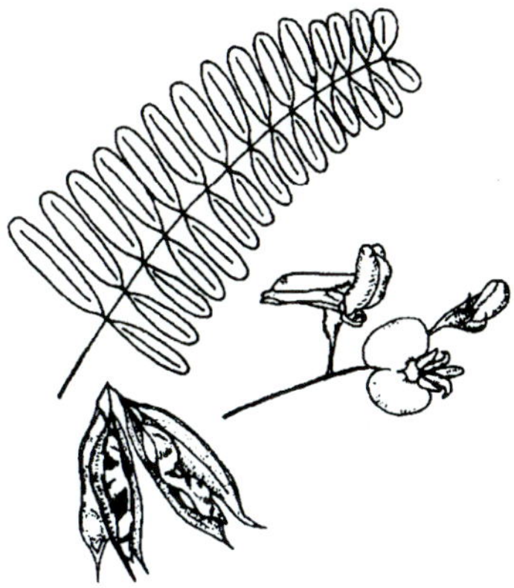

G. vesicarium

G. vesicarium

G. vesicarium

I. Diarrhea: profuse, colic; usually transient

Iris
iris
Iridaceae, iris family

Signs
- salivation, abdominal pain, profuse diarrhea $\pm$ blood
- may have oral lesions
- usually mild to moderate severity
- signs may last hours to a day

Pathology
- rare severe case; reddening, edema, ulcerations, and hemorrhages of mucosa of abomasum and small intestine

Treatment
- administer activated charcoal orally
- administer fluids and electrolytes

Disease Problem
- most rhizomatous species are toxic
- all animal species at risk, especially puppies and dogs allowed access to stored rhizomes
- possibly three irritant, hopene-type pentacyclic terpenoids, and quinines

Taxonomy and Ecology
- representative species of toxicologic concern:

I. douglasiana	Douglas's flag
I. missouriensis	western blue flag
I. pseudacorus	yellow flag, yellow water iris

- erect perennials from rhizomes, 15–150 cm tall
- flowers in terminal clusters, showy, of various colors, radially symmetrical; 3-merous
- natives distributed across much of North America; also as ornamentals

Iris cultivated

I. missouriensis

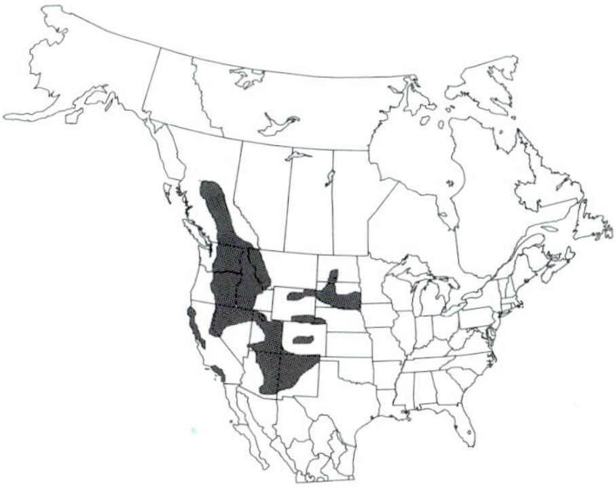

I. missouriensis

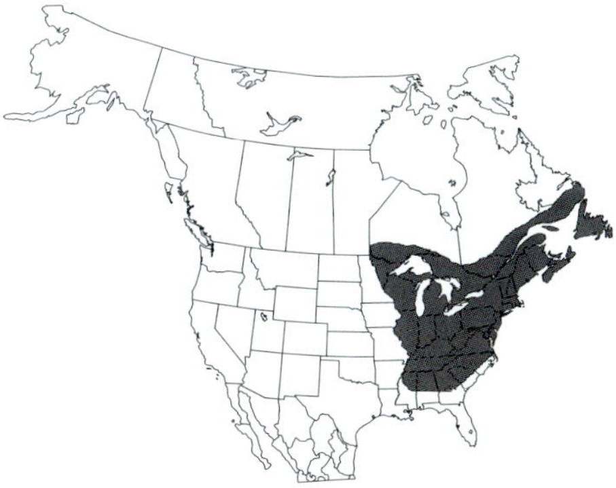

I. versicolor

I. Diarrhea: persistent, bloody and mucoid, loss of appetite; rarely cardiac effects

Kalanchoe
mother-of-millions
Crassulaceae, stonecrop family

Signs and Diagnostic Aids
- depression, loss of interest in food or water, excess salivation, mucoid and bloody diarrhea
- diarrhea may persist several days without other signs
- rare severe case: increased heart rate, labored respiration, expiratory grunting, arrhythmias, slow heart rate, heart block with dropped beats, jugular pulse, severe weakness
- increase in BUN, creatinine 2–20-fold, blood glucose 2-fold

Pathology
- reddening and some ulceration of mucosa of stomach and small intestine
- microscopically: mild myocardial necrosis

Treatment
- administer activated charcoal orally
- administer fluids and electrolytes
- atropine to increase heart rate and reverse heart block

Disease Problem
- flowers especially toxic to dogs; other household pets also at risk
- less than 1% b.w. may cause intoxication
- cardiotoxic bufadienolides in all plant parts, especially flowers
- effects cumulative
- cardiac effects rare, requires large amounts of plant

Taxonomy and Ecology
- representative species of toxicologic concern:

K. daigremontiana	mother-of-millions
K. delagonensis	mother-of-millions
K. fedtschenkoi	South American air-plant

- succulent, erect herbs, 20–100 cm tall
- leaves opposite or whorled
- flowers showy, tubular, white, green, violet, red
- introduced ornamentals; may naturalize in southern portion of continent

Kalanchoe cultivated

Kalanchoe cultivated

Kalanchoe cultivated

K. delagonensis

I. Diarrhea: with colic

Ligustrum
privet
Oleaceae, olive family

Signs
- signs variable but typically include diarrhea
- additionally abdominal pain, vomiting, ataxia
- in some instances recumbency

Pathology
- intense reddening of the mucosa of the stomach/abomasum and small intestine

Treatment
- administer activated charcoal orally
- administer fluid and electrolytes

Disease Problem
- leaves, trimmings, and berries are toxic
- probably all herbivores affected
- toxicant not definitely known
- possibly irritant, pentacyclic terpenoid glycosides, ligustrins

Taxonomy and Ecology
- representative introduced species of toxicologic concern:

L. amurense	Amur privet, Amur River privet
L. japonicum	Japanese privet, wax-leaf privet
L. lucidum	glossy privet, large-leaf privet
L. vulgare	privet, common privet, privet hedge

- deciduous or evergreen shrubs or small trees
- leaves simple, margins smooth
- flowers small, white, funnelform; petals 4
- fruits purple-black berries
- introduced ornamentals, often planted as hedges; escaping and naturalizing; pests in southeastern portion of continent

Ligustrum cultivated

Ligustrum cultivated

L. sinense

L. sinense

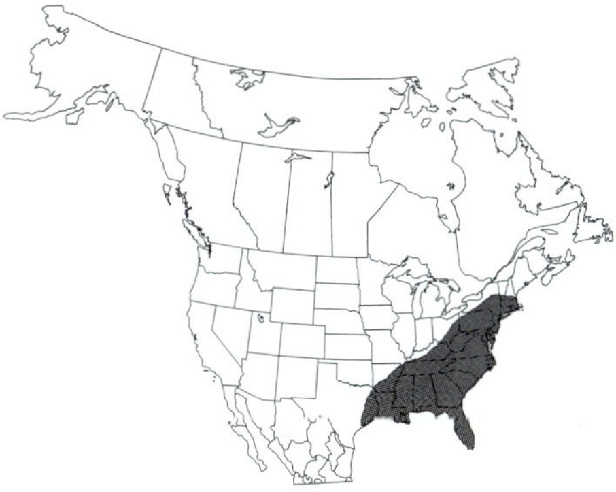

L. vulgare

I. Diarrhea: retching, excess salivation, colic

Narcissus
daffodil, jonquil
Liliaceae, lily family

Signs
- vomiting, excess salivation, restlessness, abdominal pain, diarrhea
- rarely depression, dyspnea, ataxia

Pathology
- reddening and edema of mucosa of stomach and intestines

Treatment
- administer antiemetics and antidiarrheals
- administer activated charcoal orally

Disease Problem
- bulbs, flowers, stems, raw or cooked cause digestive problem
- typically not serious
- irritant phenanthridine alkaloids; lycorine
- similar intoxication potential exists for numerous related genera of the lily family, which contain these alkaloids
- also causes contact and allergic dermatitis or rarely neurologic problems

Taxonomy and Ecology
- representative species of toxicologic concern:

N. jonquilla	jonquil
N. poeticus	poet's daffodil, poet's narcissus

- erect perennials from bulbs, 20–45 cm tall
- leaves basal, flat blades
- flowers large, showy; yellow, white, or bicolored
- introduced ornamentals

Narcissus cultivated

Narcissus

I. Diarrhea: colic, retching

Phytolacca
pokeweed
Phytolaccaceae, pokeweed family

Signs
- diarrhea in 3–4 hours in monogastrics and 6–8 hours in ruminants, lasting 1 day or more
- blood may be present in feces or color may be due to purple-red juice from berries
- pigs eating roots: ataxia, sedation, tonic seizures
- cattle: may seriously reduce milk production

Pathology
- reddening of the mucosa of stomach and small intestine, with small focal erosions

Treatment
- provide supportive care as needed
- administer fluids and electrolytes

Disease Problem
- young shoots, leaves, and seeds are typically eaten by humans without problem when cooked properly
- eating leaves (green-chop feed) or roots; transient digestive disease, monogastric animals and ruminants
- seldom overt signs of intoxication
- irritant triterpene saponins, phytolaccagenin, phytolaccinic acid, pokeberrygenin, and their respective glycosides or phytolaccosides
- also affects neuromuscular system

Taxonomy and Ecology
- one species of toxicologic concern:

P. americana	poke, pokeweed, pokeberry, inkberry, poke sallet, pigeonberry

- annual or short-lived perennial herbs from fleshy roots
- stems erect, 1–3 m tall, fleshy and purplish red at maturity
- leaves ovate to lanceolate, glabrous
- flowers in drooping or arching racemes, greenish white or cream colored
- fruits purplish black berries persisting after leaves fall
- typically associated with disturbed soil; eastern half of continent

Phytolacca americana

P. americana

P. americana

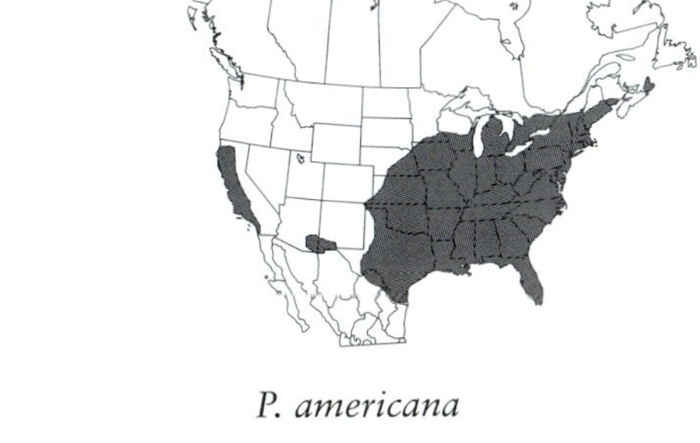

P. americana

I. Diarrhea: with blistering of mouth, colic, rarely tremors and seizures

Ranunculus
buttercup, crowfoot
Ranunculaceae, buttercup family

Signs
- erythema and swelling of muzzle, nose, lips, face
- nausea, vomiting, diarrhea, colic
- milk and urine reddish, blood tinged
- rarely: tremors, seizures, persistent weakness, dyspnea

Pathology
- edema of serosa and reddening of mucosa of abomasum and small intestine
- presence of plant fragments in ingesta

Treatment
- provide supportive care as needed

Disease Problem
- irritation of entire digestive tract may be severe
- readily eaten by sheep, even when mature and fruiting
- typically only a few animals are affected
- greater than 1% b.w. required for intoxication
- glycoside ranunculin is hydrolyzed to irritant protoanemonin when plant tissues macerated, then to inactive anemonin
- also affects urinary (hematuria) and neuromuscular systems and mammary glands (reddish and bitter milk)

Taxonomy and Ecology
- representative species of toxicologic concern:

R. abortivus	smallflower buttercup, little leaf buttercup, kidney-leaved buttercup
R. bulbosus	bulbous buttercup, common buttercup
R. sceleratus	celeryleaf buttercup, blister buttercup, ditch crowfoot
R. testiculatus	testiculate buttercup, bur buttercup, hornseed buttercup

- annual or perennial herbs, 5–100 cm tall
- leaves simple or palmately compound, alternate
- flowers solitary or in clusters, showy
- sepals 3–5, yellow or white
- petals 0–numerous, yellow, white, red, or green
- typically near water or in moist soils across continent

Ranunculus acris

Ranunculus sp.

R. acris

R. sceleratus

I. Diarrhea: severe, watery, ± blood, straining, colic, weakness

Ricinus
castor bean
Euphorbiaceae, spurge family

Signs and Diagnostic Aids

- appear 4–8 hours after well-chewed seeds eaten
- severe watery diarrhea ± blood, abdominal straining
- nausea, severe abdominal pain, decrease in blood pressure, weakness with trembling, anorexia, sweating, vomiting, sudden collapse
- copious thick, ropy salivation and hypothermia
- increase in AST, ALT, LDH, BUN, Ca, PO; decrease in blood glucose
- when leaves eaten: muscular weakness, tremors, ataxia, exercise intolerance, copious salivation

Pathology

- reddening, edema, and necrosis of mucosa of stomach and small intestine

Treatment

- administer fluids with electrolytes and glucose
- administer activated charcoal orally

Disease Problem

- seeds; hard, difficult to chew, rarely serious problem
- when well chewed, less than 0.01% b.w. may cause serious problems
- glycoprotein ricin in seed, inhibits ribosomes, prevents protein synthesis
- poorly absorbed orally; low toxicity for ruminants, greater risk for monogastric animals
- leaves and/or pericarps of fruits contain ricinin, which is much less toxic than seeds, requiring 2% b.w. for intoxication

Taxonomy and Ecology

- one polymorphic species of toxicologic concern:

R. communis castor bean, castor-oil plant, castor

- erect, almost treelike annual herb (perennial in tropics), 1–5 m tall
- leaves large, simple, alternate, palmately lobed, pungent odor
- flowers staminate and pistillate; petals lacking, pistils red/purplish red
- fruits spiny, globose; seeds ticklike, mottled
- ornamental or weedy invader in southern portion of continent

Ricinus communis

R. communis

R. communis

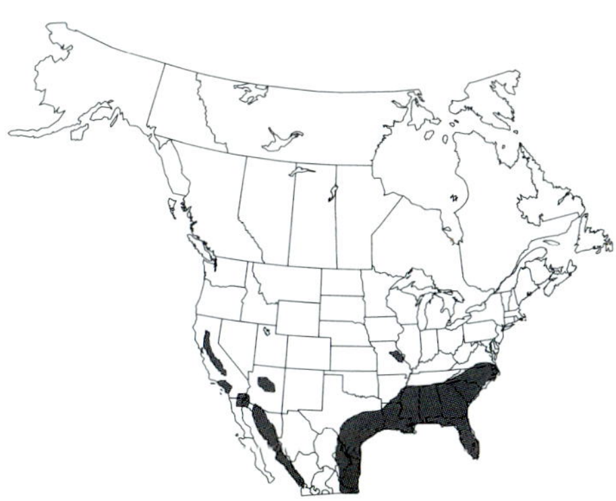

R. communis

I. Diarrhea: with straining, later perhaps muscle weakness

Senna
senna, cassia
Fabaceae, pea family

Signs and Diagnostic Aids
- diarrhea with straining within hours of ingestion
- if ingestion continues, anorexia, depression, progressive muscle weakness, eventually recumbency
- within a few days, cardiac insufficiency
- myoglobinuria, increase in serum AST, CPK

Pathology
- reddening of mucosa of stomach and small intestine
- skeletal muscles: pale, especially of pelvic limbs, with pale streaks and transversely banded stippling
- cardiac muscle: flabby with pale streaks
- microscopically: muscle swelling and vacuolation of sarcoplasm, degeneration, rupture of sarcolemmal membranes; degenerative and normal myofibres intermixed

Treatment
- provide supportive care as needed

Disease Problem
- most species cause laxative effects, self-limiting, transient
- coffee- and two-leaved sennas; skeletal and perhaps cardiac muscle degeneration
- seeds more toxic than foliage, pose serious threat for pigs and chickens when contaminate feed
- 0.1–0.3% b.w. may cause intoxication
- anthraquinones, bianthrones, anthrones; digestive effects; may also cause myopathy by inhibiting electron transport in muscle mitochondria

Taxonomy and Ecology
- four species of toxicologic concern:

S. lindheimeriana	Lindheimer's senna
S. obtusifolia	sicklepod, coffee-pod
S. occidentalis	coffeeweed, coffee-senna, wild coffee, styptic weed
S. roemeriana	two-leaved senna, twin-leaved senna

- erect, ill-scented herbs, 0.3–2 m tall
- leaves alternate, 1-pinnately compound, leaflets 1–many
- flowers in axillary racemes, petals 5, yellow or white
- legumes linear, straight or curved, brown
- waste areas or invaders of cultivated fields

Senna roemeriana

S. occidentalis

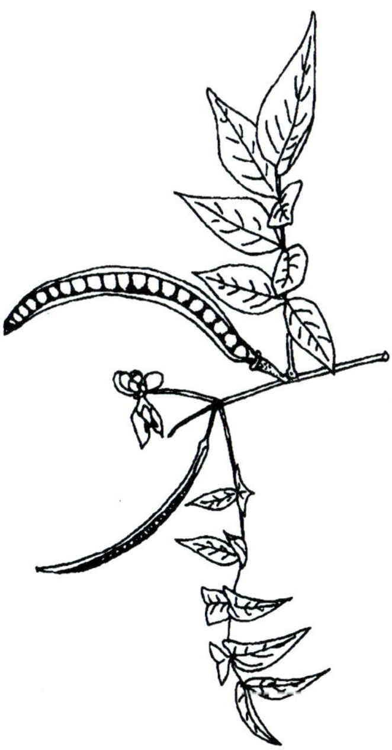

S. occidentalis

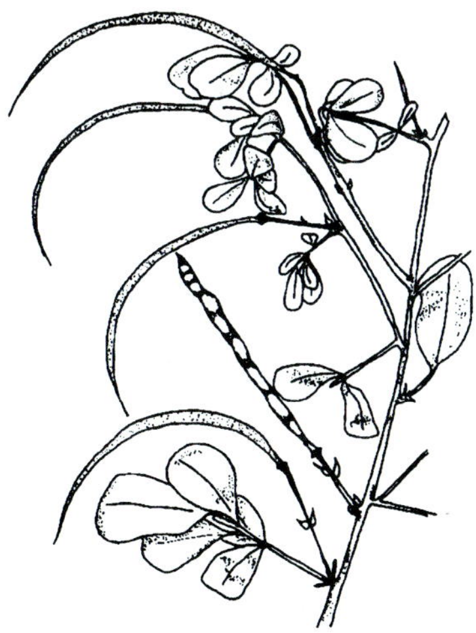

S. obtusifolia

I. Diarrhea: severe depression, weakness, colic

Sesbania
sesbania
Fabaceae, pea family

Signs and Diagnostic Aids
- depression, ruminal atony, abdominal pain, weakness, mucoid feces
- diarrhea severe, heart rate increased
- animal stands with back arched, head down, and neck extended
- neutrophilia with many immature forms

Pathology
- moderate to severe abomasal and intestinal reddening
- mild to moderate renal tubular necrosis
- presence of beans in ruminal contents

Treatment
- provide supportive care as needed
- administer activated charcoal orally
- administer fluids and electrolytes

Disease Problem
- green and ripe seeds toxic, foliage little risk
- 0.2–1% b.w. may cause intoxication
- sesbanimides primary toxicants; saponins and sesbanine may contribute; inhibition of smooth muscle and vasodilation

Taxonomy and Ecology
- two species of toxicologic concern:

S. drummondii	poison bean, coffee bean, rattlebush, Drummond rattlebush, rattlebox
S. punicea	purple sesban, purple rattlebush, purple rattlebox, daubentonia

- shrubs or small trees to 10 m tall
- leaves 1-pinnately compound, lack terminal leaflets
- flowers pealike, yellow, dull scarlet, purple, white
- legumes elongated, 4-winged

Sesbania drummondii

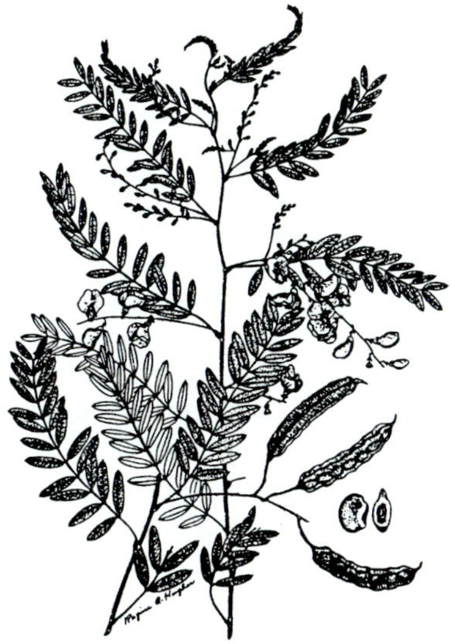

S. punicea

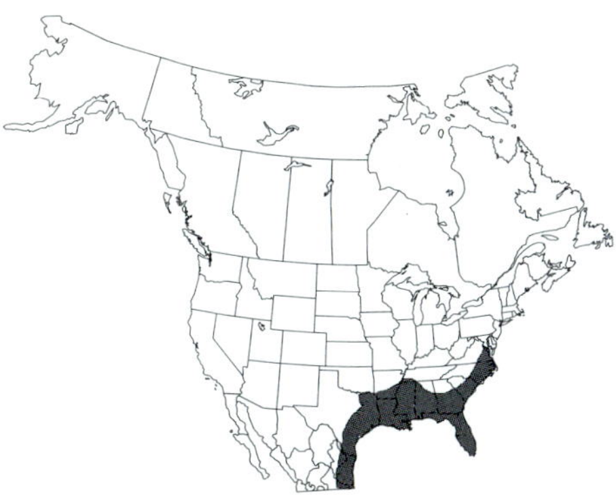

S. drummondii

II. Impaction of Rumen or Stomach: severe colic with lack of gut motility, rapid heart rate

Diospyros
persimmon
Ebenaceae, ebony family

Signs and Diagnostic Aids
- severe colic, absence of gut motility, elevated heart rate especially in horses
- leukocytosis
- perforation of gastric ulcerations with signs of peritonitis

Pathology
- presence of large mass of pulp, skins, seeds in stomach
- reddening and perhaps ulceration of stomach mucosa

Treatment
- oral administration of mineral oil
- mechanical breakup of bezoars with an endoscope, or dissolution by cellulose or papain
- surgical intervention to remove mass
- administer low molecular weight polyethylene glycol orally to bind tannins in early stages of disease

Disease Problem
- fruits and seeds, congeal into obstructing masses, termed phytobezoars, in stomach and/or small intestine
- fruits most hazardous when eaten prior to ripening before first frosts
- soluble condensed tannins, cross-link in stomach, form congealed mass of fruit, seeds
- astringent action on stomach mucosa may result in irritation and ulceration

Taxonomy and Ecology
- two native and one cultivated species of toxicologic concern:

D. kaki	Japanese persimmon, kaki
D. texana	Texas persimmon, black persimmon
D. virginiana	common persimmon, possumwood

- small trees, solitary or colonial
- leaves simple, alternate
- flowers staminate and pistillate
- petals 4, yellowish green, creamy white, or orange
- fruits berries, yellow, pale orange, red-orange, or black
- open woodlands in southern part of continent; cultivated in warmer areas

Diospyros virginiana

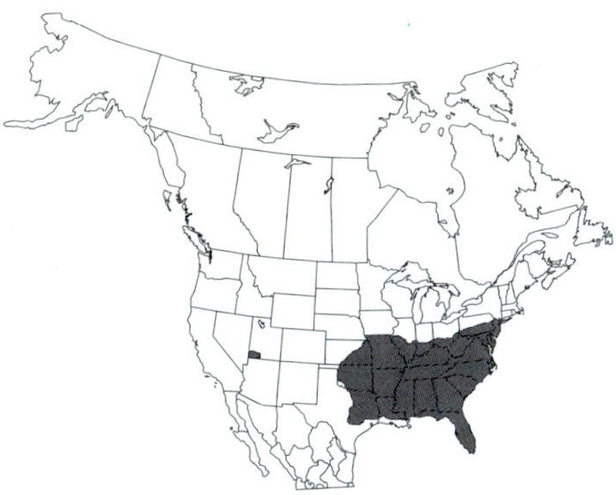

D. virginiana

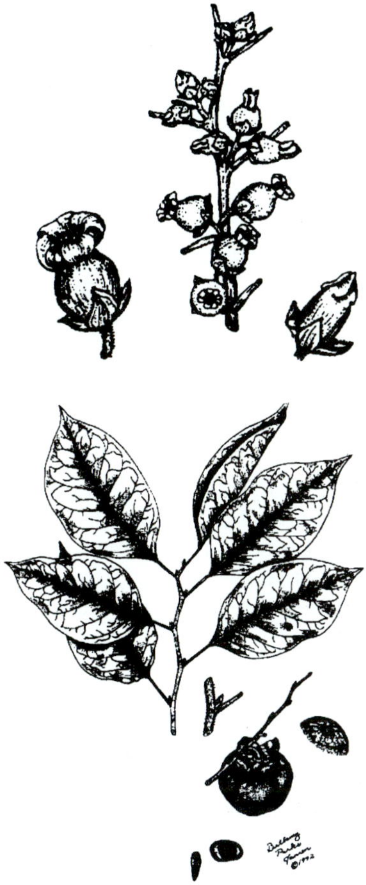

D. virginiana

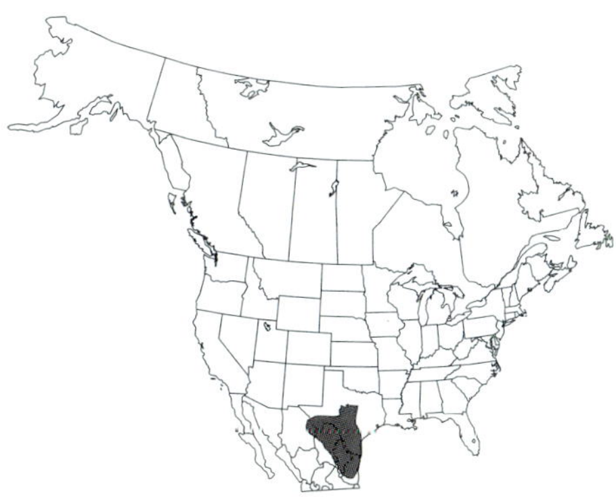

D. texana

IV. Salivation: regurgitation with greenish saliva and nasal discharge, green muzzle, repeated swallowing movements, coughing

Helenium
sneezeweed
Asteraceae, sunflower family

Signs

- uncommon acute disease: abrupt onset, depression, decreased appetite, ruminal atony, mild bloat, stand with arched back, walk with difficulty
- serous nasal discharge, cough, wheezing with green saliva, green stains of muzzle, repeated swallowing
- more common subacute disease: gradual deterioration, weight loss, dehydration, regurgitation, foamy green saliva
- death due to dehydration and starvation

Pathology

- no distinctive lesions

Treatment and Prevention

- provide general supportive care
- anticholinergic, antidepressant, and antiarrhythmic drugs provide only temporary relief
- prevention: possibly supplement with sulfur proteins, sulfur amino acids or antioxidants (ethoxyquin)
- prevention: to insure selection of nontoxic forage plants, use open, quiet herding, slow grazing, uniform range use, careful selection of bed grounds
- prevention: combine grazing of cattle, sheep, goats

Disease Problem

- metabolic derangement, digestive disturbance, neurologic and cardiotoxic effects
- sesquiterpene lactones, helenanolides; helenalin, neohelenalin
- highest concentrations in flowers
- 0.25–2% b.w. may cause intoxication depending on plant species; *H. microcephalum* is most toxic
- tenulin in *H. amarum* causes bitter milk and meat
- inhibition of enzymes; aerobic and anaerobic metabolism, phosphofructokinase, hexokinase, succinic dehydrogenase
- cumulative effects sometimes over several weeks
- similar disease produced by *Baileya* and *Psilostrophe*

Taxonomy and Ecology

- representative species of toxicologic concern:

H. amarum	bitter sneezeweed, Spanish daisy
H. autumnale	common sneezeweed, autumn sneezeweed, sneezewort
H. flexuosum	purplehead sneezeweed
H. hoopesii	orange sneezeweed
H. microcephalum	smallhead sneezeweed

- herbs, some aromatic, some with winged stems, 10–150 cm tall
- leaves simple, alternate, of various shapes
- flowers (florets) borne in numerous, showy, small heads
- ray florets 3-lobed, yellow, orange, or reddish brown; disk florets yellow, brown, or purple
- variety of habitats, mainly central and eastern portion of continent

Helenium flexuosum

H. amarum

H. amarum

H. autumnale

IV. Salivation: regurgitation with greenish saliva and nasal discharge, green muzzle, repeated swallowing movements, coughing

Hymenoxys
bitterweed
Asteraceae, sunflower family

Signs

- uncommon acute disease: abrupt onset, depression, decreased appetite, ruminal atony, mild bloat, stand with arched back, walk with difficulty
- serous nasal discharge, cough, wheezing with green saliva, green stains of muzzle, repeated swallowing
- more common subacute disease: gradual deterioration, weight loss, dehydration, regurgitation, foamy green saliva
- death due to dehydration and starvation

Pathology

- no distinctive lesions

Treatment and Prevention

- provide general supportive care
- anticholinergic, antidepressant, and antiarrhythmic drugs provide only temporary relief
- prevention: possibly supplement with sulfur proteins, sulfur amino acids, or antioxidants (ethoxyquin)
- prevention: control plant growth or prevent exposure of animals to plant
- prevention: combined grazing with cattle, sheep, goats

Disease Problem

- metabolic derangement; digestive disturbance; neurologic and cardiotoxic effects
- dihemiacetal pseudoguaianolides; hymenoxon, hymenoxynin, hymenolide, vermeerin
- concentrations high in both foliage and flowers
- 0.5–1.5% b.w. may cause intoxication
- inhibition of enzymes; aerobic and anaerobic metabolism, phosphofructokinase, hexokinase, succinic dehydrogenase
- cumulative action over several weeks
- also affects neuromuscular and respiratory systems

Taxonomy and Ecology

- four species of toxicologic concern:

H. lemmonii	alkali hymenoxys, Lemmon's hymenoxys
H. odorata	bitterweed, bitter rubberweed
H. richardsonii	Colorado rubberweed, pingue
H. subintegra	Arizona rubberweed

- erect or ascending, aromatic herbs, with resinous exudate, 10–60 cm tall
- leaves simple, alternate, of various shapes
- flowers (florets) borne in numerous, showy, small heads
- ray florets 3-lobed, yellow; disk flowers yellow
- western portions of continent

Hymenoxys odorata

H. odorata

H. odorata

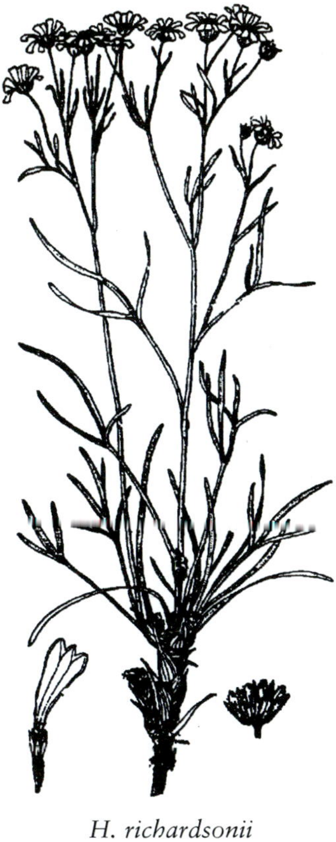

H. richardsonii

V. Salivation: profuse, with mild inflammation of mouth and lips

Dieffenbachia
dumb cane
Araceae, arum family

Signs

- immediate swelling of lips, tongue, mouth
- profuse salivation, rarely vomiting and diarrhea
- typically mild, occasionally severe with swelling and pain rendering animal unable to eat for several days
- conjunctivitis
- effects rarely extend to esophagus, stomach
- serious intoxications unlikely because rapid onset of pain and swelling terminates animal's interest in plant

Treatment

- generally not needed
- rare severe cases: sedate, feed soft foods, control pain and inflammation with nonsteroidal anti-inflammatory drugs
- apply steroid ointment for conjunctivitis

Disease Problem

- houseplant; pets and small children at greatest risk
- mechanical irritation to mucous membranes
- packets of grooved, needlelike calcium oxalate crystals (raphides) in idioblast cells of leaves and stems
- raphides extruded from idioblasts when mucilage in cell swells or when external pressure applied
- crystals persist in cornea for several days

Taxonomy and Ecology

- two ornamental species of toxicologic concern:

D. maculata	dumb cane
D. seguine	dumb cane, American arum, poison arum

- herbs with stout, erect stems, to 2.5 m tall
- conspicuous leaf scars
- leaves simple, often variegated
- seldom flower indoors

Dieffenbachia cultivated

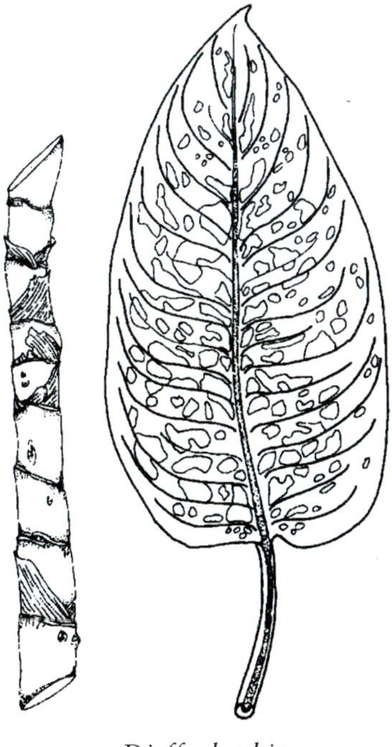

Dieffenbachia

V. Salivation: profuse, with mild inflammation of mouth and lips

Epipremnum
golden pothos, devil's ivy
Araceae, arum family

Signs
- immediate swelling of lips, tongue, mouth
- profuse salivation, rarely vomiting and diarrhea
- typically mild, occasionally severe with swelling and pain rendering animal unable to eat for several days
- conjunctivitis
- effects rarely extend to esophagus and stomach
- serious intoxications are unlikely because rapid onset of pain and swelling terminates animal's interest in plant

Treatment
- generally not needed
- rare severe cases: sedate, feed soft foods, control pain and inflammation with nonsteroidal anti-inflammatory drugs
- apply steroid ointment for conjunctivitis

Disease Problem
- houseplant; pets and small children at greatest risk
- mechanical irritation to mucous membranes
- packets of grooved, needlelike calcium oxalate crystals (raphides) in idioblast cells of leaves and stems
- raphides extruded from idioblasts when mucilage in cell swells or when external pressure applied
- crystals persist in cornea for several days

Taxonomy and Ecology
- one ornamental species of toxicologic concern:

E. aureum ivy arum, devil's ivy, golden pothos

- herbaceous vine with adventitious roots from stems
- leaves simple, often variegated
- seldom flower indoors
- perhaps most common introduced houseplant

Epipremnum cultivated

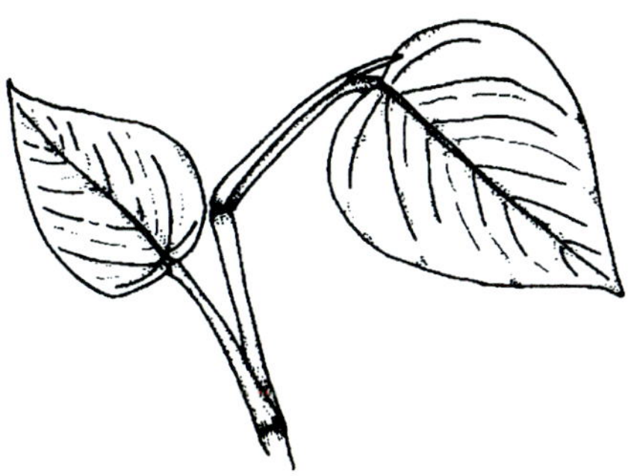

Epipremnum

V. Salivation: profuse, with mild inflammation of mouth and lips

Philodendron
philodendron
Araceae, arum family

Signs

- immediate swelling of lips, tongue, mouth
- profuse salivation; rarely vomiting and diarrhea
- typically mild, occasionally severe with swelling and pain rendering the animal unable to eat for several days
- conjunctivitis
- effects rarely extend to esophagus, stomach
- serious intoxications are unlikely because rapid onset of pain and swelling terminates animal's interest in plant

Treatment

- generally not needed
- rare severe cases: sedate, feed soft foods, control pain and inflammation with nonsteroidal anti-inflammatory drugs
- apply steroid ointment for conjunctivitis

Disease Problem

- houseplant; pets and small children at greatest risk
- mechanical irritation to mucous membranes
- packets of grooved, needlelike calcium oxalate crystals (raphides) in idioblast cells of leaves and stems
- raphides extruded from idioblasts when mucilage in cell swells or when external pressure applied
- crystals present in cornea for several days
- rarely renal effects

Taxonomy and Ecology

- numerous cultivars of different species
- climbing, evergreen herbs or shrubs
- leaves simple of varying shapes
- seldom flower indoors

Philodendron cultivated

Philodendron cultivated

Philodendron

VI. Salivation: conspicuously profuse, excess lacrimation, diarrhea, stiffness

Trifolium
clover
Fabaceae, pea family

Signs
- profuse salivation, stringy, clear, viscid saliva hanging from mouth
- hypothermia, lacrimation, diarrhea, bloat
- increased frequency of urination, stiffness, inappetance
- marked decrease in milk production
- loss of weight and condition
- often transient, not life threatening

Pathology
- not a lethal disease

Treatment
- administer atropine for partial alleviation of signs
- removal of infected forage usually followed by prompt amelioration of signs

Disease Problem
- "slobbers disease," associated primarily with red clover
- infestation of clover by fungus *Rhizoctonia leguminicola*, black patch disease characterized by dark splotches on leaves
- ingestion of large amounts of fresh forage, hay, or silage
- cholinergic alkaloid, slaframine or rhizotoxin, bioactivated by liver
- *Trifolium* also affects skin (photosensitization) and reproductive system (dysfunction due to estrogenic activity)

Taxonomy and Ecology
- cultivated perennials of toxicologic concern:

T. hybridum	alsike clover
T. incarnatum	crimson clover
T. pratense	red clover, purple clover
T. repens	white clover, Dutch clover, ladino clover
T. subterraneum	subterranean clover

- erect or creeping herbs
- leaves palmately compound, leaflets 3
- flowers pealike white, red, pink, yellow, or purplish

Trifolium incarnatum cultivated

T. pratense

T. pratense

VII. Salivation: profuse, persistent retching/vomiting, colic, repeated swallowing, vocal indications of pain

Kalmia
sheep laurel
Ericaceae, heath family

Signs
- anorexia, profuse salivation, repeated swallowing, persistent retching/vomiting, bloat, severe abdominal pain, irregular respiration
- regurgitation or vomiting may be projectile
- bellowing or other indications of pain
- acute effects last 24 hrs, weakness or neurologic effects 2–3 days
- severe cases: weakness, reluctance or inability to stand
- rare cardiac effects may result in decrease in heart rate and arrhythmias such as AV block

Pathology
- no distinctive lesions
- possible presence of leaves in the stomach

Treatment
- no specific antidotes, relief of symptoms
- administer atropine
- administer activated charcoal orally

Disease Problem
- diterpenoid grayanotoxins in all parts of plants
- bind sodium channels, slow opening and closing, decrease ion selectivity, especially in excitable cells; neurologic, cardiac, muscular effects
- increased reflex activity, increased vagal influence
- decreased blood pressure and heart rate (arrhythmias)
- similar disease produced by other members of family: *Andromeda*, *Ledum*, *Leucothoe*, *Lyonia*, *Menziesia*, and *Pieris*

Taxonomy and Ecology
- four species of toxicologic concern:

K. angustifolia	sheep laurel, lambkill
K. latifolia	mountain laurel, laurel, calico-bush, sheep laurel, poison laurel
K. microphylla	small-leaved laurel, alpine laurel bog laurel, western swamp laurel
K. polifolia	swamp laurel, bog laurel

- evergreen or deciduous shrubs or small trees
- leaves simple, leathery, margins entire
- flowers showy, wheel- or bell-shaped
- petals 5, fused, white, pink, light rose, or rose-purple
- mainly in moist acidic soils of mountain areas

Kalmia latifolia

K. latifolia

K. latifolia

K. microphylla

VII. Salivation: profuse, persistent retching/vomiting, colic, repeated swallowing, vocal indications of pain

Rhododendron
rhododendron, azalea
Ericaceae, heath family

Signs
- anorexia, profuse salivation, repeated swallowing, persistent retching/vomiting, bloat, severe abdominal pain, irregular respiration
- regurgitation or vomiting may be projectile
- bellowing or other indications of pain
- acute effects last 24 hrs, weakness or neurologic effects 2–3 days
- severe cases: weakness, reluctance or inability to stand
- rare cardiac effects may result in decrease in heart rate and arrhythmias such as AV block

Pathology
- no distinctive lesions
- presence of leathery leaves in stomach

Treatment
- no specific antidotes, relief of symptoms
- administer atropine
- administer activated charcoal orally

Disease Problem
- diterpenoid grayanotoxins in all parts of plants
- bind sodium channels, slow opening and closing, decrease ion selectivity, especially in excitable cells; neurologic, cardiac, muscular effects
- increased reflex activity, increased vagal influence, decreased blood pressure and heart rate (arrhythmias)
- rarely intoxications may occur due to contamination of honey by bees feeding on *Rhododendron*
- similar disease produced by other members of family: *Andromeda*, *Ledum*, *Leucothoe*, *Lyonia*, *Menziesia*, and *Pieris*

Taxonomy and Ecology
- representative species of toxicologic concern:

R. albiflorum	white rhododendron, Cascade rhododendron
R. catawbiense	mountain rosebay, rosebay laurel, catawba rhododendron
R. macrophyllum	Pacific rhododendron, western rhododendron, California rosebay
R. maximum	rosebay, white laurel
R. occidentale	western azalea

- deciduous species called azaleas, evergreen taxa called rhododendrons, with a few exceptions
- evergreen or deciduous shrubs or small trees
- leaves simple, alternate, margins revolute
- flowers showy, rotate or campanulate or funnelform
- petals 5, fused, of various colors except blue
- moist woodlands of eastern and western portions of continent; some cultivated as ornamentals

Rhododendron catawbiense

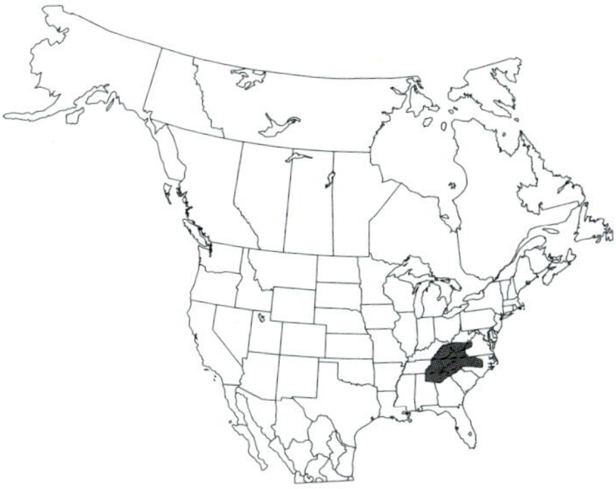

R. catawbiense

R. maximum

R. albiflorum

VII. Salivation: profuse, ropy strings of saliva hanging from mouth, retching/vomiting, colic, severe depression

Zigadenus
death camas
Liliaceae, lily family

Signs
- abrupt onset of profuse frothy salivation (ropy strings hanging from mouth)
- depression, nausea, regurgitation, retching, abdominal pain, grinding of teeth
- weakness severe, trembling, animals prefer to lie down, rise only with difficulty
- sheep may become hyperexcitable despite weakness
- signs similar to those of organophosphate intoxication

Pathology
- no distinctive lesions
- presence of plant fragments in rumen
- analyze ruminal contents for alkaloids

Treatment
- administer activated charcoal orally
- administer atropine or sympathomimetics for cardiovascular effects

Disease Problem
- mainly sheep; 0.1–1% b.w. may cause intoxication
- recovery in 1–2 days typical in cattle, horses
- risk is greatest in early spring when plants are most likely to be eaten
- toxic effects; digestive, hypotension, neurologic
- cevanine-type *Veratrum* azasteroid alkaloids; mono- and polyesters of germine, protoverine, zygadenine
- increase reflex activity, sensitize afferent pathway receptors, increasing and extending negative afterpotential resulting in repetitive discharge in response to single stimuli in nerves and muscle

Taxonomy and Ecology
- representative species of toxicologic concern:

Z. nuttallii	Nuttall's death camas
Z. paniculatus	foothill death camas, sand-corn
Z. venenosus	meadow death camas

- erect herbs from dark brown/black bulbs or thick rhizomes, 20–120 cm tall
- leaves basal, long
- flowers in terminal racemes or panicles, white to greenish yellow, sometimes purple-tinged
- variety of habitats across continent, common in West

Zigadenus nuttallii

Z. nuttallii

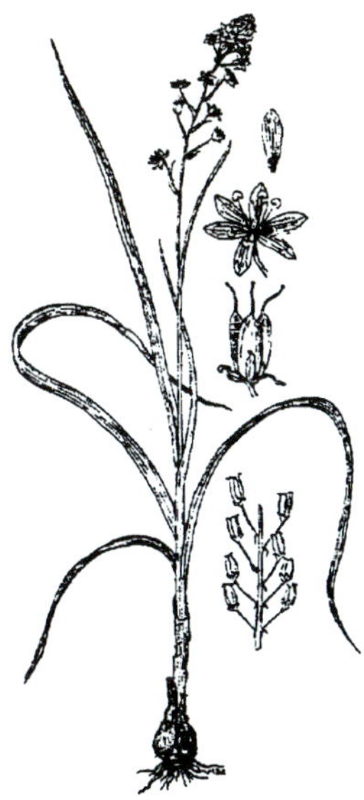

Z. venenosus

Z. nuttallii

IX. Weight Loss: conspicuous, mainly cattle in hot weather, poor appearance, rough haircoat, decreased food intake

Festuca
fescue
Poaceae, grass family

Signs and Diagnostic Aids

- summer-slump in weight gain; especially when temperatures exceed 32°C
- increased body temperature (0.5°C) and respiratory rate; decreased food intake, growth, and milk production; and reproductive effects
- cattle: poor appearance, rough haircoat that does not shed, stand in shade and/or water instead of eating
- fat necrosis: firm masses in adipose tissue of abdominal cavity, with occasional vascular or luminal obstruction of intestine or urogenital system
- diagnosis: detection of fungal endophyte in plant tissues, stain to identify fungal elements, culture of seed tissues, use ELISA tests for fungal elements
- additional problem of "fescue-foot" syndrome; winter and early spring; swelling, reddening of coronary band(s) with clear line of demarcation around leg just above fetlock, dry gangrene and loss of foot and tail switch
- syndrome same as for "ergotism"

Pathology

- no lesions for summer-slump syndrome

Treatment and Prevention

- supplement with other forages
- increase initial stocking rates to keep fescue at low height and reduce access to more toxic leaf sheaths
- peripheral vasoconstrictive effects subside within 1 week after removal from toxin
- protection by thiamine (1 g/day), ivermectin alone or with implants of progesterone/estradiol benzoate

Disease Problem

- cattle: "summer-slump" or "summer fescue syndrome"
- decreased feed intake and weight gain
- ergot-type alkaloids; endophytic fungus, *Neotyphodium coenophialum*, entirely within stems, leaf sheaths, spikelets; does not alter appearance of plant
- ergopeptides and ergoclavine alkaloids; highest in caryopses/seeds, leaf sheaths (loss of toxicity after 2 years)
- also causes reproductive effects and possibly increased risk of laminitis in horses
- fescue-foot is erratic in occurrence
- similar syndrome possible with other forages; *Lolium*, *Cyperus*

Taxonomy and Ecology

- three cultivated and lawn species of toxicologic concern:

F. arundinacea	tall fescue, reed fescue
F. pratensis	meadow fescue, English bluegrass
F. rubra	chewings fescue, red fescue

- perennial bunchgrasses with membranous ligules, 0.5–1.5 m tall
- inflorescences panicles
- spikelets laterally compressed, florets 2–35, glumes shorter than lemmas

F. arundinacea

Festuca arundinacea

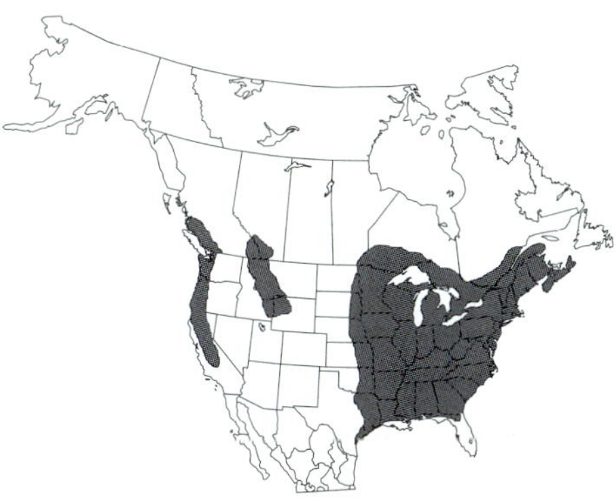

F. arundinacea

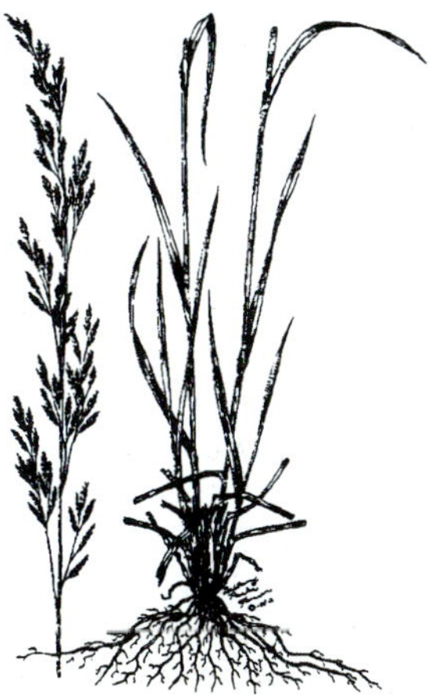

F. arundinacea

IX. Weight Loss: conspicuous; ruminants, mainly diminished weight gain cattle

Kochia
kochia
Chenopodiaceae, goosefoot family

Signs and Diagnostic Aids
- diminished weight gain in most animals after grazing for several weeks
- severe cases: impairment of appetite, depression, swollen eyelids, crusty areas of nasal pad, erythema and epithelial sloughing of nonpigmented areas of skin (photosensitization)
- icterus, depression, weakness, diarrhea, dehydration
- mild to moderate increase in serum GGT, AST, LDH, SDH, CPK, AP; increase in direct and indirect bilirubin

Pathology
- skin lesions characteristic of photosensitization
- kidneys slightly swollen, pale
- microscopically: diffuse renal tubular dilation, foamy, flattened, or necrotic epithelial cells; mild to moderate hepatocellular swelling, bile stasis, mild necrosis and/or fibrosis

Treatment
- provide general nursing care
- prevent animal access to plants

Disease Problem
- useful forage but overall value somewhat ambiguous
- adverse effects most likely when *Kochia* composes more than 50% of diet or is eaten for several months
- possibly due to oleanolic acid genin saponins
- also sporadically causes polioencephalomalacia (PEM) (page 160)

Taxonomy and Ecology
- one introduced species of toxicologic concern:

K. scoparia	burning bush, summer cypress, fireball, fireweed, Mexican fireweed, kochia

- herbaceous annual from taproot, to 2 m tall
- stems profusely branched, purplish red at maturity
- leaves simple, alternate, narrow
- flowers small, in axillary leaflike bracts at branch ends
- common in disturbed soils; naturalized across western half of continent; cultivated as a forage

Kochia scoparia

K. scoparia

K. scoparia

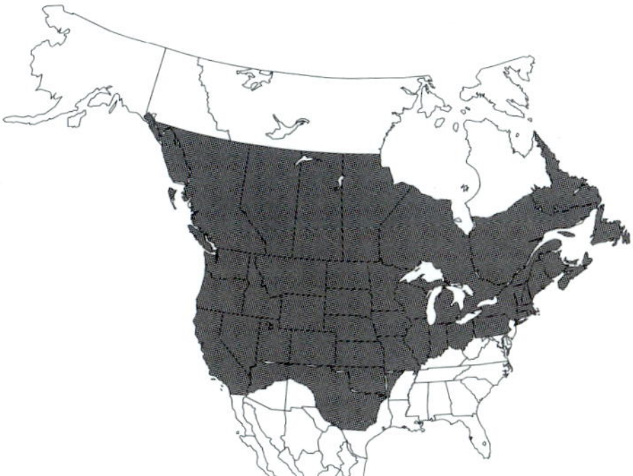

K. scoparia

4
Plants Affecting the Eyes

With the exception of those taxa, such as members of the arum family, that produce compounds that cause mechanical irritation, relatively few toxic plants have a primary effect on the eyes. Most notable, however, is *Datura*, thorn apple, and other members of the Solanaceae. An additional 11 genera in 9 families do affect the eye secondarily.

Principal Signs

I. **Blindness:** apparent or actual, without obvious signs of mechanical irritation
II. **Blindness:** with neurologic signs of polioencephalomalacia (PEM)
III. **Blindness:** due to third-eyelid protrusion
IV. **Mechanical Irritation:** reddening, tearing
V. **Mydriasis:** plus other anticholinergic effects; dry mouth

Plants Causing Problems

I. **Blindness:** apparent or actual, without obvious signs of mechanical irritation

Secondary	
Astragalus	milk vetch (page 130)
Brassica	mustard (page 14)
Pteridium	bracken fern (page 8, 138)
Zanthoxylum	prickly ash (page 158)
Tertiary	
Descurania	tansy mustard

II. **Blindness:** with neurologic signs of polioencephalomalacia (PEM)

Secondary	
Astragalus	locoweed (page 128)
Brassica	mustard (page 14, 160)
Kochia	kochia (page 86, 160)
Phalaris	canarygrass (page 202)

Tertiary

Conyza	horseweed

III. Blindness: due to third-eyelid protrusion

Secondary

Calycanthus	sweet shrub (page 172)
Conium	poison hemlock (page 204)
Nicotiana	tobacco (page 210)

IV. Mechanical Irritation: reddening and tearing

Secondary

Dieffenbachia	dumb cane (page 70)
Epipremnum	devil's ivy (page 72)
Philodendron	philodendron (page 74)

Tertiary

Alocasia	elephant's ear
Arisaema	jack-in-the-pulpit
Calla	water arum
Capsicum	pepper
Euphorbia	spurge
Lysichiton	skunk cabbage
Monstera	Swiss cheese plant
Symplocarpus	skunk cabbage
Zantedeschia	calla lily

V. Mydriasis: plus other anticholinergic effects; dry mouth

Primary

Datura	thorn apple

Secondary

Solanum	nightshade (page 142)

Tertiary

Brugmansia	angel's trumpet
Hyoscyamus	henbane
Solandra	chalice vine

V. Mydriasis: marked and persistent; photophobia, dry mucous membranes, and other anticholinergic effects

Datura
thorn apple
Solanaceae, nightshade family

Signs and Diagnostic Aids
- persistent mydriasis, photophobia
- restlessness, irritability, weakness, sinus tachycardia, dryness of mucous membranes, thirst, constipation, increased body temperature
- rarely: ataxia, paresis, later depression, respiratory failure
- presence of tropane alkaloids in urine or stomach contents; confirm by placing drop of animal's urine in a test animal's eye and observing mydriasis

Pathology
- no distinct lesions present

Treatment
- most cases do not require treatment
- administer activated charcoal orally
- may counteract with cholinergic stimulants such as physostigmine

Disease Problem
- intoxications may result from ingestion of seeds contaminating feed or from fresh foliage
- toxic dosage is 0.1% b.w.: plants extremely unpalatable—unless fresh foliage is eaten rapidly, disease in livestock is typically self-limiting due to reduced motility of gut and decreased appetite
- anticholinergic tropane alkaloids: l-hyoscyamine, scopolamine (l-hyoscine); atropine d-,l-hyoscyamine
- competitive antagonism of acetylcholine at muscarinic cholinergic receptors

Taxonomy and Ecology
- representative species of toxicologic concern:

D. inoxia	large-flowered thorn apple, sacred datura, Indian apple
D. metel	black datura, devil's trumpet
D. stramonium	jimsonweed, Jamestown weed, thorn apple
D. wrightii	sacred datura, Indian apple, tolguacha

- erect herbs or subshrubs, 0.5–1.5 m tall
- leaves simple, alternate, short petiolate
- flowers large, showy, solitary in branch forks
- corollas tubular to trumpet shaped, white, yellow, violet to purple
- fruits prickly capsules
- cultivated as ornamentals; others naturalized in waste areas across continent

Datura stramonium

D. stramonium

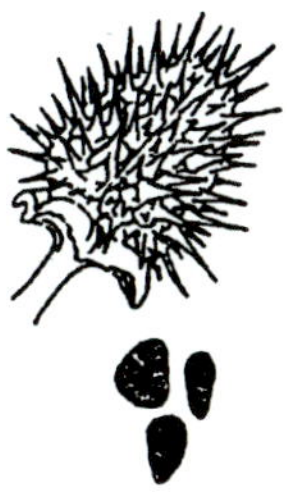

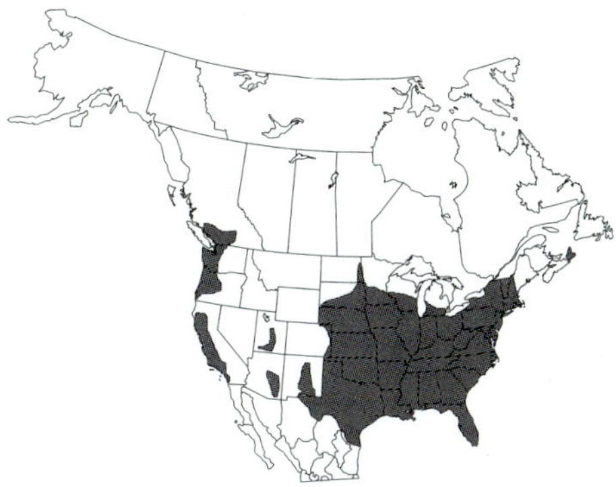

D. stramonium

D. stramonium

D. wrightii

5
Plants Affecting the Liver

As the body's principal site of metabolic biotransformation, the liver is subjected to the effects of numerous toxic compounds produced by plants. It is remarkably resilient to these compounds, their precursors, and their breakdown products. Its function and structural organization, however, is altered specifically by seven genera in six families. An additional eight genera also produce changes in addition to their primary effects on other organ systems. Although the clinical signs of icterus, dark urine, and bile-stained feces may indicate that the liver has been definitely affected by these compounds, confirmation of toxic effects is often dependent upon postmortem observations.

Principal Signs

I. **Icterus:** dark urine, sometimes head pressing and photosensitization; liver necrosis/fibrosis
II. **Loss of Weight and Condition:** neurologic signs especially in horses, tenesmus, rarely photosensitization, bile-stained feces, depression; portal liver fibrosis/ductile proliferation mainly; "pyrrolizidine alkalosis"
III. **Photosensitization:** weakness, dark urine, weight loss, icterus; crystalloid cholangiohepatopathy

Plants Causing Problems

I. **Icterus:** dark urine, sometimes head pressing and photosensitization; liver necrosis/fibrosis

Primary	
Cycas	sago palm
Lantana	lantana
Phyllanthus	leaf-flower
Tetradymia	horsebrush
Secondary	
Conium	poison hemlock (page 204)
Eupatorium	white snakeroot (page 206)

Gossypium	cotton seed (page 32)
Lolium	perennial ryegrass (page 198)
Xanthium	cocklebur (page 186)
Tertiary	
Cestrum	green cestrum
Indigofera	indigo
Karwinskia	coyotillo
Psathyrotes	desert velvet
Trifolium	clover
Zamia	coontie

II. Loss of Weight and Condition: neurologic signs especially in horses, tenesmus, rarely photosensitization, bile-stained feces, depression; portal liver fibrosis/ductile proliferation mainly; "pyrrolizidine alkalosis"

Primary	
Crotalaria	rattlepod
Cynoglossum	hound's tongue
Senecio	groundsel
Tertiary	
Amsinckia	fiddleneck
Echium	bugloss
Heliotropium	heliotrope

III. Photosensitization: weakness, dark urine, weight loss, icterus; crystalloid cholangiohepatopathy

Secondary	
Agave	lechuguilla (page 256)
Nolina	sacahuista (page 258)
Panicum	panicum (page 260)
Tribulus	goathead (page 262)

I. Icterus: dark urine, persistent episodic vomiting, anorexia; liver necrosis/fibrosis; dogs

Cycas
sago palm
Cycadaceae, cycad family

Signs and Diagnostic Aids
- dogs mainly at risk
- persistent vomiting, with periodic episodes sometimes lasting for hours
- excess salivation, increased water consumption
- anorexia, diarrhea or constipation, icterus
- decrease in platelets; increase in serum bilirubin (especially conjugated), hepatic enzymes, BUN, creatinine
- neurologic effects in cattle rare in North America

Pathology
- liver enlarged, nutmeg
- microscopically: centrilobular hepatic necrosis, fibrosis, bile duct proliferation, bile stasis

Treatment
- provide supportive care

Disease Problem
- seeds, leaves, and pulp of cones cause hepatotoxic effects especially in dogs
- two types of toxins
- cycasin, macrozamin: glycosides of methylazoxymethanol (MAM), hydrolyzed in digestive tract to hepatotoxic aglycone MAM that alkylates DNA and RNA
- neurotoxic amino acid, β-*N*-methylamino-L-alanine (L-BMAA)
- also affects digestive and neuromuscular systems
- *Zamia*, coontie, in Florida is similarly toxic

Taxonomy and Ecology
- two introduced species of toxicologic concern:

C. circinalis	crozier cycad, fern palm, false sago palm
C. revoluta	sago-cycas, sago palm, Australian nut palm

- perennial evergreens; stems unbranched, 2–10 m tall
- leaves large, 1-pinnately compound foliage
- plants unisexual; pollen and seed large
- cultivated as outdoor ornamentals in frost-free areas and as tub plants elsewhere

Cycas revoluta cultivated

C. revoluta

C. revoluta cultivated

I. Icterus: dark urine, sometimes head pressing and photosensitization, severe depression, anorexia; liver necrosis/fibrosis

Lantana
lantana
Verbenaceae, vervain family

Signs and Diagnostic Aids

- depression, loss of appetite, marked decrease in ruminal motility, severe constipation
- second or third day: icterus, photosensitization, diarrhea ± blood
- increase in serum bilirubin (especially conjugated), SDH, AST, slower clearance of BSP
- death due to hepatic and renal failure
- subacute to chronic disease is mainly associated with photosensitization

Pathology

- liver swollen and orange-colored due to bile staining
- gall bladder much enlarged with edematous walls
- kidneys swollen accompanied by perirenal edema
- reddening and edema of gut mucosa
- microscopically: portal fibrosis, bile duct hyperplasia, mild to moderate centrilobular necrosis

Treatment

- administer activated charcoal or bentonite orally
- administer fluids, glucose, and electrolytes
- sodium thiosulfate, 0.5 g/kg b.w. i.v. of possible benefit

Disease Problem

- leaves toxic but not ripe fruits and seeds
- all animal species at risk
- 1% b.w. may be sufficient to cause intoxication
- pentacyclic triterpenoid lantadenes, biotransformed in liver, excreted in bile; damage bile canalicular membranes and microvilli, decrease ATPase activity causing closure and blockage of bile flow; obstructive cholangitis
- irritation of gut mucosa, constipation or diarrhea
- photosensitization presumably due to phylloerythrin

Taxonomy and Ecology

- two species of toxicologic concern:

L. camara	lantana, red sage, yellow sage
L. montevidensis	creeping lantana, trailing lantana, weeping lantana

- perennial herbs or shrubs, foliage malodorous, stems 4-sided, to 2 m tall
- leaves opposite or whorled with toothed margins
- flowers in dense spikes or heads; petals 4 or 5, white, red, yellow, orange, blue, or purple
- fruits globose, fleshy, glossy black drupes
- ornamentals or naturalized weeds in warmer southern areas

Lantana camara

L. camara

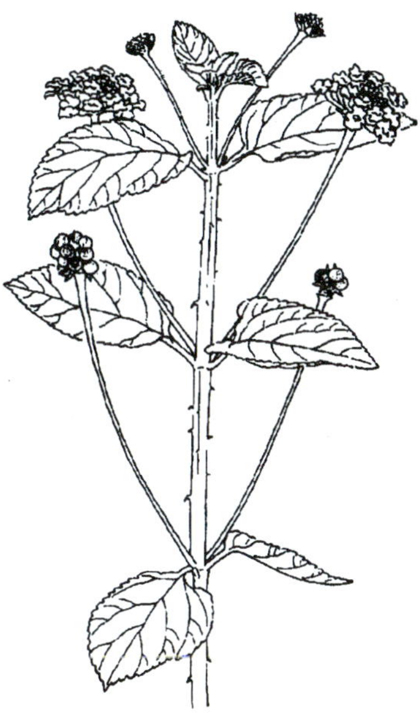

L. camara

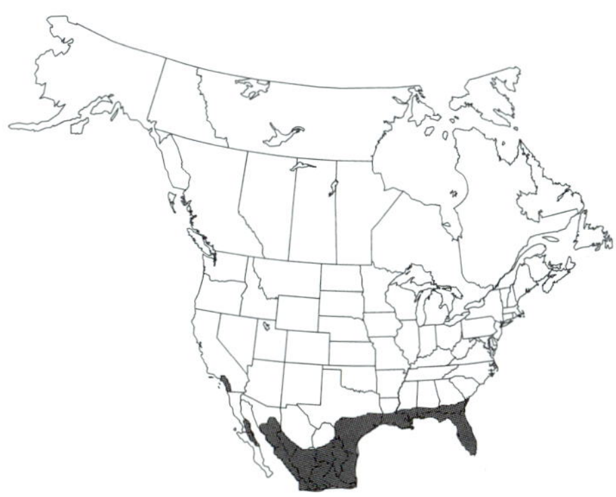

L. camara

I. Icterus: dark urine, listlessness, loss of condition, clay-colored diarrhea, tenesmus, sometimes head pressing and photosensitization; liver necrosis and fibrosis

Phyllanthus
leaf flower
Euphorbiaceae, spurge family

Signs
- listlessness, anorexia, loss of weight and condition
- diarrhea clay-colored
- tenesmus, sometimes leading to rectal prolapse
- icterus, photosensitization, incessant walking

Pathology
- liver yellow, swollen, or small and firm
- edema of mesentery and gall bladder wall
- microscopically: hepatocellular necrosis and bile duct proliferation

Treatment
- general supportive care

Disease Problem
- mainly cattle, also sheep, goats, and possibly horses
- extensively grazed; 1–2% b.w. fresh foliage produces intoxication
- acute to chronic hepatic degeneration
- specific toxicant unknown; affects nursing calves
- also affects skin and digestive system

Taxonomy and Ecology
- one species of known toxicologic risk:

P. abnormis leaf flower

- annual herb with milky sap, 10–50 cm tall
- leaves simple, alternate, 2-ranked, with stipules
- flowers axillary, paired at nodes, small, lacking petals
- fruits spherical, often reddish
- occurs in dry, sandy soils of Oklahoma, Texas, Florida

Phyllanthus abnormis

P. abnormis

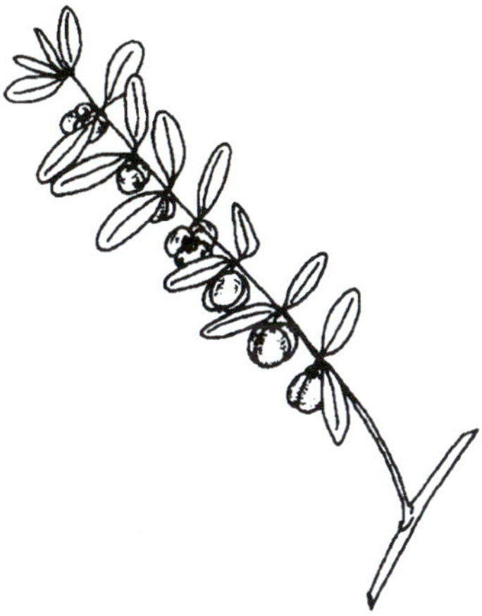

P. abnormis

P. abnormis

I. Icterus: dark urine, anorexia, depression, sometimes head pressing and photosensitization; liver necrosis and fibrosis

Tetradymia
horsebrush
Asteraceae, sunflower family

Signs and Diagnostic Aids
- abrupt onset; loss of appetite, lethargy/depression, uneasiness, tremors, ataxia, labored respiration
- possibly head pressing, incessant walking, icterus
- photosensitization may develop in some animals
- serum AST, SDH, and GGT increased 10–20 times
- BSP clearance time increased ten times

Pathology
- liver tan, friable
- gall bladder distended with bile
- microscopically: marked centrilobular hepatic necrosis

Treatment
- provide general nursing care
- limiting intake of green forage may reduce severity of signs of photosensitization

Disease Problem
- mainly sheep; "bighead"
- due to ingestion of both *Artemisia* and *Tetradymia*
- all growth stages, both fresh and dry, toxic
- the toxic dosage ranges from 0.5 to 1.0% b.w., higher dosage leading to more acute effects
- *T. glabrata* in early spring causes severe liver disease and death
- *T. canescens* more likely to cause photosensitization
- tetradymol and other furanoeremophilane derivatives
- prior consumption of *Artemisia* enhances hepatic biotransformation of tetradymol to toxicant
- liver unable to clear phylloerythrin

Taxonomy and Ecology
- two species of greatest toxicologic concern:

T. canescens	spineless horsebrush, common horsebrush, little gray horsebrush
T. glabrata	lizard shade, littleleaf horsebrush

- profusely branched shrubs, 10–150 cm tall
- leaves simple, alternate
- flowers (florets) borne in cylindrical heads
- ray florets absent
- disk florets 4–8, yellow to cream
- common; different vegetation types; intermountain region; Canada to Baja California

Tetradymia canescens

T. canescens

T. canescens

T. canescens

T. glabrata

II. Loss of Weight and Condition: neurologic signs especially in horses, tenesmus, bile-stained feces, depression, ataxia, persistent walking, circling, pushing on objects, rarely photosensitization; portal liver fibrosis/ductile proliferation mainly; "pyrrolizidine alkalosis"

Crotalaria
rattlebox
Fabaceae, pea family

Signs
- initial signs: poor appetite and weight loss
- days/weeks later: ataxia, icterus, stupor, severe weakness
- horses: sleepy, drag rear legs when walking, have episodes of unmanageable behavior, restless, self-mutilation, persistent walking or circling, leaning/pushing against solid objects
- cattle: similar signs, also depression, excess salivation, ataxia, bloat, icterus, tenesmus with partial eversion of rectum, diarrhea sometimes tarlike
- pigs: hair loss, stunted, unthrifty, loss of appetite, fail to gain weight, lethargy, labored respiration

Pathology
- liver firm, slightly shrunken, grayish blue, yellow, or tan (purple in more acute), mottled with irregular reddish-brown foci ("hobnail or nutmeg liver")
- viscera edematous
- cattle: gall bladder distended with bile, walls edematous, abundant ascitic fluid
- microscopically: mild hepatic necrosis, inter- and intralobular fibrosis, bile duct proliferation, hepatocellular megalocytosis and karyomegaly, central vein fibrosis, hyperplastic nodules

Treatment
- provide general nursing care, sedation, parenteral fluids
- provide high-quality and high-energy feeds, high-quality protein; molasses, corn, sorghum, grass hay, or beet pulp
- supplements of sulfur amino acids of questionable value
- efforts to increase hepatic glutathione content helpful

Disease Problem
- horses: "river bottom disease," now rare
- toxins 11-membered, macrocyclic "closed" diester pyrrolizidine alkaloids (PAs); monocrotaline
- PAs metabolized in liver to highly reactive pyrroles; alkylate DNA, alter RNA and protein synthesis, impair cell division, thus cells grow slowly but are unable to divide, cell and nucleus become very large (hepatocytomegaly and karyomegaly)
- renal, minor pulmonary effects in pigs, cattle, poultry

Taxonomy and Ecology
- representative species of toxicological concern:

C. sagittalis	rattlebox, rattlepod, crotalaria
C. spectabilis	showy crotalaria

- annual herbs
- leaves simple, palmately compound, leaflets 3, stipules present or absent
- flowers pealike, yellow
- fruits inflated legumes
- occurs in eastern half of continent, especially southeast

Crotalaria spectabilis

C. sagittalis

C. sagittalis

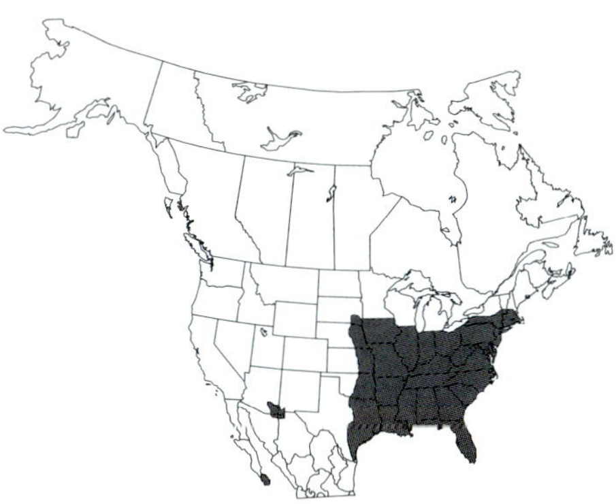

C. sagittalis

II. Loss of Weight and Condition: neurologic signs especially in horses, tenesmus, bile-stained feces, depression, head pressing, circling, aimless pacing, rarely photosensitization; portal liver fibrosis/ductile proliferation mainly; "pyrrolizidine alkalosis"

Cynoglossum
hound's tongue
Boraginaceae, forget-me-not family

Signs and Diagnostic Aids

- poor body condition, weight loss, icterus, red urine (Hb)
- abrupt onset of head pressing, aimless pacing, ataxia, chewing, yawning, intermittent drowsiness, tenesmus, either constipation or diarrhea
- rarely photosensitization
- degenerative left shift; increase in serum AST, GGT, AP
- presence of sulfur-conjugated pyrroles in blood or liver

Pathology

- liver firm, slightly shrunken, grayish blue, yellow, or tan (purple in more acute), mottled with irregular reddish-brown foci ("hobnail or nutmeg liver")
- viscera edematous
- cattle: gall bladder distended with bile, walls edematous, abundant ascitic fluid
- microscopically: mild hepatic necrosis, inter- and intralobular fibrosis, bile duct proliferation, hepatocellular megalocytosis and karyomegaly, central vein fibrosis, hyperplastic nodules

Treatment

- provide general nursing care, sedation, parenteral fluids
- provide high-quality and high-energy feeds, high-quality protein; molasses, corn, sorghum, grass hay, or beet pulp
- supplements with sulfur amino acids of questionable value
- efforts to increase hepatic glutathione content helpful

Disease Problem

- mainly contaminated hay; fresh plants seldom eaten
- the toxic dosage is high but it may be spread out over a long period with only a few plants eaten per day
- increasing in geographical range and abundance as an intoxication risk
- open diesters (heliosupine), several monoester PAs
- high levels in leaves early in growing season of first- or second-year plants

Taxonomy and Ecology

- one species of recognized toxicologic concern:

C. officinale	hound's tongue, beggars-lice, dog bur, dog's tongue, stick-tight

- erect herbs, 30–120 cm tall
- leaves basal and/or on stem
- flowers blue, borne along axis shaped like neck of fiddle
- fruits prickly nutlets in 4s, stick to hair, wool, clothing
- naturalized as a serious weed problem throughout most of North America; native species occur in Pacific Coast states and British Columbia

Cynoglossum officinale

C. officinale

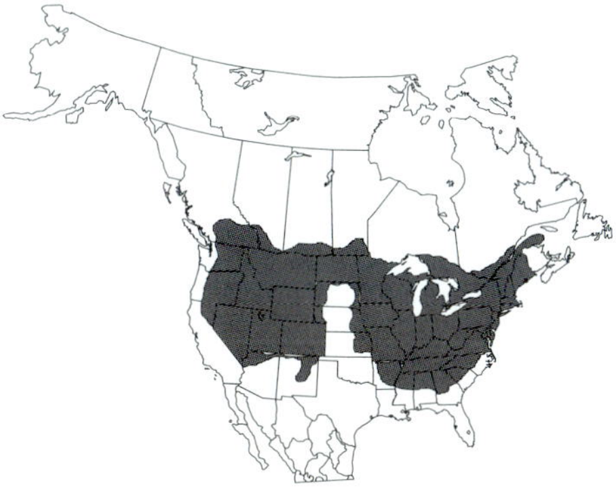

C. officinale

C. officinale

II. Loss of Weight and Condition: neurologic signs especially in horses, tenesmus, bile-stained feces, depression, head pressing, circling, aimless and/or incessant walking, rarely photosensitization; portal liver fibrosis/ductile proliferation mainly; "pyrrolizidine alkalosis"

Senecio
groundsel
Asteraceae, sunflower family

Signs and Diagnostic Aids

- depression, loss of appetite and weight
- horses: abrupt onset of head pressing, incessant or aimless walking or pacing, persistent chewing of fences, yawning, drowsiness, tenesmus, ataxia, ascites, and diarrhea or constipation
- sometimes increased thirst; Hb in urine
- cattle: similar but slower onset, tenesmus, rectal prolapse, less-pronounced neurologic signs
- increase in GLDH, GGT, AP, and AST; smaller increase in SDH, LDH
- decrease in rate of serum BSP clearance
- increase in serum bile acids; less than 50 umol/l indicates a poor prognosis
- test for pyrroles in plant, liver, or blood

Pathology

- liver firm, slightly shrunken, grayish blue, yellow, or tan (purple in more acute), mottled with irregular reddish-brown foci ("hobnail or nutmeg liver")
- viscera edematous; omentum thick, bluish gray
- cattle: gall bladder distended with bile, walls edematous, abundant ascitic fluid
- microscopically: mild hepatic necrosis, inter- and intralobular fibrosis, bile duct proliferation, hepatocellular megalocytosis and karyomegaly, central vein fibrosis, hyperplastic nodules

Treatment

- provide general nursing care, sedation, parenteral fluids
- provide high-quality and high-energy feeds, high-quality protein; molasses, corn, sorghum, grass hay, or beet pulp
- supplements of sulfur amino acids of questionable value
- efforts to increase hepatic glutathione content helpful

Disease Problem

- risk mainly with plant species in areas with long growing seasons or those locally plentiful and/or producing abundant herbage
- horses and cattle most often affected; sheep and goats much more resistant
- toxic dosage typically 25–50% b.w. ingested over several months
- flowering plants are greatest risk
- delay in onset of signs; may be triggered by stress
- pyrrolizidine alkaloids (PAs), 12-membered macrocyclic diesters, jacobine, retrorsine, riddelline, senecionine alkylate DNA, alter RNA and protein synthesis
- also affects urinary system

Taxonomy and Ecology

- some prairie/woodland species now classified as *Packera*
- representative species posing greatest risk:

S. flaccidus	threadleaf groundsel, woolly groundsel
S. jacobaea	tansy ragwort, stinking willy

S. riddellii	sand groundsel, Riddell's groundsel
S. vulgaris	common groundsel

- herbs or shrubs, 10–130 cm tall
- leaves simple, alternate
- flowers (florets) borne in numerous, showy heads
- ray florets yellow, red, purple, or white; rarely absent
- disk florets typically yellow, rarely white or purple
- toxic species mainly in western North America

Senecio jacobaea

S. flaccidus

S. riddellii

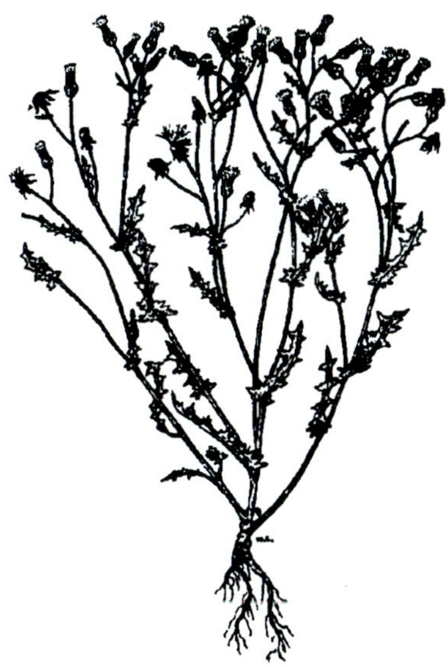

S. vulgaris

6
Plants Affecting the Neuromuscular System

As might be expected, the neuromuscular system of any animal is quite sensitive to the presence of toxic compounds produced by plants. Clinical signs indicating such alterations may appear rapidly and be dramatic—seizures, ataxia, paralysis—or be gradual in their onset and more subtle, for example, those of depression or muscle weakness. It is difficult to assign plants to one specific type of sign since there may be a gradation of effects from ataxia to tremors to seizures depending on severity and duration of disease. Thus, plants may be cross-listed among the various categories. We have attempted placement based on the predominant type of effect. Changes in the neuromuscular system also commonly reflect secondarily changes in other organs and organ systems induced by plant toxins. Forty-five genera in 14 families directly affect the neuromuscular system. An additional 67 genera produce significant changes in the system in addition to their primary effects on other systems, and many of the toxicants produce multiple suites of clinical signs.

Principal Signs

I. **Ataxia or Incoordination**
II. **Depression or Somnolence:** disorientation, drowsiness
III. **Laryngeal Paresis**
IV. **Limb Abnormalities:** thoracic limb flexion, hypermetria, high-stepping gait
V. **Limb Abnormalities:** pelvic limb flexion, exaggerated flexion, hopping gait, or knuckling of fetlocks
VI. **Limb Abnormalities:** paresis/paralysis, muscle weakness, recumbency
VII. **Limb Abnormalities:** rigidity, stiffness
VIII. **Lip and Tongue Movement Abnormalities:** difficulty in eating and swallowing
IX. **Pacing, Head Pressing, Circling, Chewing, Yawning**
X. **Seizures:** usually abrupt onset
XI. **Seizures:** abrupt onset and sudden death
XII. **Tremors:** beginning with neck and shoulder muscles: ears erect, nervous, stiff, wobbly, swaying on pelvic limbs
XIII. **Tremors and Generalized Tetany**
XIV. **Violent Behavior**

Plants Causing Problems

I. Ataxia or Incoordination

Primary

Acacia	guajillo
Aesculus	buckeye
Artemisia	sagebrush
Astragalus	locoweed
Astragalus	milk vetch
Laburnum	golden-chain tree
Lathyrus	pea
Oxytropis	locoweed
Pteridium	bracken fern

Secondary

Amaranthus	pigweed (page 270
Astrolepis	jimmy fern (page 166)
Brassica	mustard, rape, kale (page 14)
Conium	poison hemlock (page 204)
Corydalis	corydalis (page 176)
Cycas	sago palm (page 98)
Cynodon	Bermudagrass (page 196)
Datura	thorn apple (page 92)
Delphinium	larkspur (page 148)
Gelsemium	jessamine (page 178)
Halogeton	halogeton (page 278)
Helenium	sneezeweed (page 66)
Hymenoxys	bitterweed (page 68)
Kallstroemia	caltrop (page 146)
Karwinskia	coyotillo (page 150)
Kochia	kochia (page 86, 160)
Ligustrum	privet (page 50)
Lupinus	lupine (page 180, 230)
Melia	chinaberry (page 182)
Paspalum	paspalum (page 200)
Phalaris	canarygrass (page 202)
Phytolacca	pokeweed (page 54)
Senna	senna (page 60)
Solanum	nightshade (page 142)
Sorghum	sudan (page 18, 190)
Veratrum	false hellabore (page 232)
Xanthium	cocklebur (page 186)
Zigadenus	death camas (page 82)

Tertiary

Aloysia	white brush
Anagallis	scarlet pimpernel
Conyza	horseweed
Indigofera	indigo

Lobelia	lobelia
Lycopersicon	tomato
Papaver	Icelandic poppy
Peganum	African rue
Rumex	sorrel
Solidago	goldenrod
Zamia	coontie

II. Depression or Somnolence: disorientation, drowsiness

Primary	
Cannabis	marijuana
Solanum	nightshade
Secondary	
Centaurea	star thistle, knapweed (page 154)
Conium	poison hemlock (page 204)
Eupatorium	white snakeroot (page 206)
Isocoma	goldenbush (page 208)
Nicotiana	tobacco (page 210)
Tertiary	
Baileya	desert marigold
Lobelia	lobelia
Lycopersicon	tomato
Papaver	poppy

III. Laryngeal Paresis

Secondary	
Lathyrus	pea (page 134)

IV. Limb Abnormalities: thoracic limb flexion, hypermetria, high-stepping gait

Secondary	
Aesculus	buckeye (page 124)
Astragalus	locoweed (page 128, 220)
Karwinskia	coyotillo (page 150)
Lathyrus	pea (page 134)
Lupinus	lupine (page 180, 230)
Oxytropis	locoweed (page 136)
Zanthoxylum	prickly ash (page 158)

V. Limb Abnormalities: pelvic limb flexion, exaggerated flexion, hopping gait, or knuckling of fetlocks

Primary	
Hypochaeris	cat's ear
Kallstroemia	caltrop

Secondary

Astragalus	milk vetch (page 130)
Laburnum	golden-chain tree (page 132)
Tribulus	puncture vine (page 262)

VI. Limb Abnormalities: paresis/paralysis/muscle weakness, recumbency

Primary	
Delphinium	larkspur
Karwinskia	coyotillo
Thermopsis	golden banner
Secondary	
Astragalus	milk vetch (page 130)
Cynodon	Bermudagrass (page 196)
Gelsemium	jessamine (page 178)
Paspalum	paspalum (page 200)
Senna	senna (page 60)
Sorghum	sudan (page 18, 190)
Tertiary	
Aconitum	monkshood
Hilaria	tobosagrass
Lobelia	lobelia
Robinia	black locust

VII. Limb Abnormalities: stiffness, rigidity

Secondary	
Aesculus	buckeye (page 124)
Astragalus	milk vetch (page 130)
Astrolepis	jimmy fern (page 166)
Baccharis	baccharis (page 168)
Calycanthus	sweet shrub (page 172)
Conium	poison hemlock (page 204)
Corydalis	corydalis (page 176)
Delphinium	larkspur (page 148)
Eupatorium	white snakeroot (page 206)
Gelsemium	jessamine (page 178)
Helenium	sneezeweed (page 66)
Hymenoxys	bitterweed (page 68)
Isocoma	goldenbush (page 208)
Karwinskia	coyotillo (page 150)
Lathyrus	pea (page 134)
Sophora	mescal bean (page 184)
Tertiary	
Peganum	African rue
Robinia	black locust

VIII. Lip and Tongue Movement Abnormalities: difficulty in eating and swallowing

Primary	
Centaurea	starthistle, knapweed
Prosopis	mesquite
Zanthoxylum	prickly ash
Secondary	
Astragalus	locoweed (page 128, 220)
Eupatorium	white snakeroot (page 206)
Gelsemium	jessamine (page 178)
Tertiary	
Descurainia	tansy mustard

IX. Pacing, Head Pressing, Circling, Chewing, Yawning

Primary	
Kochia (PEM)	kochia (page 86, 160)
Secondary	
Astragalus	locoweed (page 128, 220)
Brassica	mustard, rape, kale(page 14, 160)
Crotalaria	rattlebox (page 106)
Cynoglossum	hound's tongue (page 108)
Eupatorium	white snakeroot (page 206)
Lathyrus	pea (page 134)
Lupinus	lupine (page 180, 230)
Phalaris	canarygrass (page 202)
Senecio	groundsel (page 110)
Tetradymia	horsebrush (page 104)
Tertiary	
Amsinckia	fiddleneck
Conyza	horseweed
Echium	blueweed
Heliotropium	heliotrope
Indigofera	indigo

X. Seizures: usually abrupt onset

Primary	
Albizia	mimosa tree
Asclepias	milkweed
Astrolepis	jimmy fern
Baccharis	baccharis
Brunfelsia	yesterday-today-tomorrow
Calycanthus	sweet shrub
Cicuta	water hemlock
Corydalis	corydalis
Gelsemium	jessamine

Lupinus	lupine
Melia	chinaberry
Sophora	mescal bean
Xanthium	cocklebur
Secondary	
Amaranthus	pigweed (page 270)
Astragalus	locoweed (page 128, 220)
Brassica	mustard (page 14, 160)
Kallstroemia	caltrop (page 146)
Kochia	kochia (page 86, 160)
Laburnum	golden-rain tree (page 132)
Lathyrus	pea (page 134)
Lupinus	lupine (page 180, 230)
Ranunculus	buttercup (page 56)
Solanum	nightshade (page 142)
Sorghum	sorghum (page 18, 190)
Tertiary	
Conyza	horseweed
Drymaria	drymary
Indigofera	indigo
Solidago	goldenrod

XI. Seizures: abrupt onset with sudden death

Primary	
Prunus	cherry
Sorghum	sorghum
Suckleya	poison suckleya
Triglochin	arrowgrass
Tertiary	
Acacia	acacia
Amelanchier	serviceberry
Cercocarpus	mountain mahogany
Cynodon	stargrass
Glyceria	mannagrass
Heteromeles	Christmas berry
Sambucus	elderberry
Sorghastrum	Indian grass
Stillingia	stillingia

XII. Tremors: beginning with neck and shoulder muscles: ears erect, nervous, stiff, wobbly, swaying on pelvic limbs

Primary	
Cynodon	Bermuda grass
Lolium	perennial ryegrass
Paspalum	paspalum
Phalaris	canarygrass

Secondary	
Sophora	mescal bean (page 184)
Tertiary	
Hilaria	tobosagrass
Malva	mallow
Solidago	goldenrod

XIII. Tetany and Generalized Tremors

Primary	
Conium	poison hemlock
Eupatorium	white snakeroot
Isocoma	goldenbush
Nicotiana	tobacco
Secondary	
Aesculus	buckeye (page 124)
Albizia	mimosa tree (page 162)
Astragalus	milk vetch (page 130)
Baccharis	baccharis (page 168)
Brunfelsia	yesterday-today-tomorrow (page 170)
Calycanthus	sweet shrub (page 172)
Cannabis	marijuana (page 140)
Delphinium	larkspur (page 148)
Eupatorium	white snakeroot (page 206)
Halogeton	halogeton (page 278)
Helenium	sneezeweed (page 66)
Hymenoxys	bitterweed (page 68)
Isocoma	goldenbush (page 208)
Karwinskia	coyotillo (page 150)
Laburnum	golden-rain tree (page 132)
Pteridium	bracken fern (page 8, 138)
Ranunculus	buttercup (page 56)
Ricinus	castor bean (page 58)
Sarcobatus	black greasewood (page 280)
Solanum	nightshade (page 142)
Thermopsis	golden banner (page 152)
Tertiary	
Aconitum	monkshood
Baileya	desert marigold
Bassia	bassia
Descurainia	tansy mustard
Dicentra	staggerweed
Oxalis	wood sorrel
Peganum	African rue
Psilostrophe	paperflower
Rumex	sorrel
Sartwellia	sartwellia

XIV. Violent Behavior

Secondary

Artemisia	sagebrush (page 126)
Astragalus	locoweed (page 128)
Cicuta	water hemlock (page 174)
Laburnum	golden-rain tree (page 132)
Oxytropis	locoweed (page 136)
Panicum	panicum (page 260)

I. Ataxia or Incoordination: peculiar wobbly movement on pelvic limbs, weight loss, placement problem with thoracic limbs

Acacia
guajillo
Fabaceae, pea family

Signs
- weight loss despite reasonable appetite
- ataxia; unique, as if pelvic limbs and joints were rubber; wobbly gait, sideways movements
- proprioceptive deficit in thoracic limbs
- recumbent animals remain alert with reasonable appetite
- may have reproductive dysfunction and possibly decreased pregnancy rates

Pathology
- no distinctive lesions

Treatment
- generally reversible if removed from pasture and fed a high-quality diet

Disease Problem
- "guajillo wobbles" or "limberleg," sheep and goats
- ruminants subsisting on leaves and pods for 6–9 months
- risk primarily during dry periods when grass and other forages not available
- goats at high risk as browsers
- probable cause: sympathetic amines such as tyramine, phenethylamine, and hordenine

Taxonomy and Ecology
- one species of definite toxicologic concern:

A. berlandieri guajillo

- shrub or small tree
- leaves 2-pinnately compound, terminal leaflets absent
- flowers borne in spherical clusters, cream, white, sweet scented
- legumes flat, irregularly constricted
- abundant on limestone ridges and plains throughout southern Texas and adjacent Mexico

Acacia berlandieri

A. berlandieri

A. berlandieri

I. Ataxia or Incoordination: staggering gait, trembling, sawhorse stance, muscle spasms, muscular rigidity

Aesculus
buckeye
Hippocastanaceae, horse chestnut family

Signs
- sawhorse stance and a reluctance to stand or move
- ataxia, uneasy staggering gait, hypermetria, trembling, hyperesthesia
- seizures with muscle spasms, postural abnormalities, extensor rigidity, collapse
- self-limiting, lasting perhaps 12–48 hours

Pathology
- no significant lesions
- presence of seeds or fragments in rumen

Treatment
- rarely needed, toxic effect decreases plant intake

Disease Problem
- disease in spring from foliage and buds; in fall from seeds
- most serious intoxications due to *A. hippocastanum*, introduced horse chestnut
- dosage of 0.5–1.0% b.w. may result in intoxication
- mixture of pentacyclic oleanane triterpenoids, called saponins when complexed as glycosides with glucose and glucuronic acid; escigenin, barringtogenol-C, hippocaesculin, and protoescigenin
- increased sensitivity of ruminants possibly due to increased hydrolysis of glycosides releasing presumed neurotoxic aglycone in reducing environment of rumen

Taxonomy and Ecology
- representative species of known toxicologic concern:

A. californica	California buckeye
A. flava	sweet buckeye, yellow buckeye
A. glabra	western buckeye, white buckeye, Ohio buckeye, Texas buckeye, fetid buckeye
A. hippocastanum	horse chestnut
A. pavia	red buckeye

- deciduous shrubs or small trees, 2–6 m tall, except for *A. hippocastanum* up to 30 m tall
- leaves palmately compound, opposite leaflets, 5–11
- flowers in panicles
- petals 4–5, free, clawed, white, greenish yellow, or red
- fruits capsules
- seeds 1–3, light orange or dark brown
- rocky and sandy soils, in woods, thickets, fencerows, and along streams; also ornamentals

Aesculus glabra

A. glabra

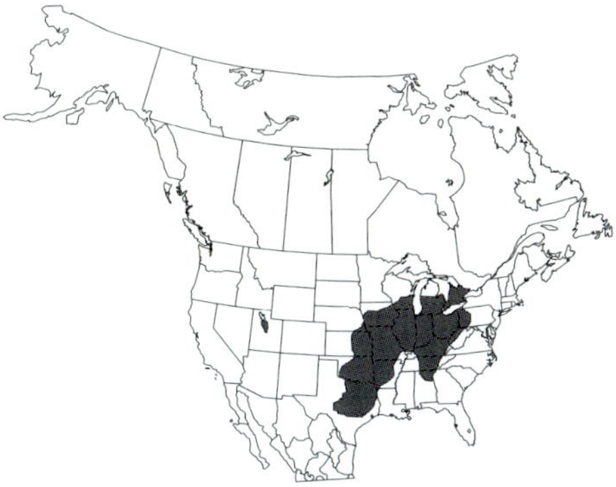

A. glabra

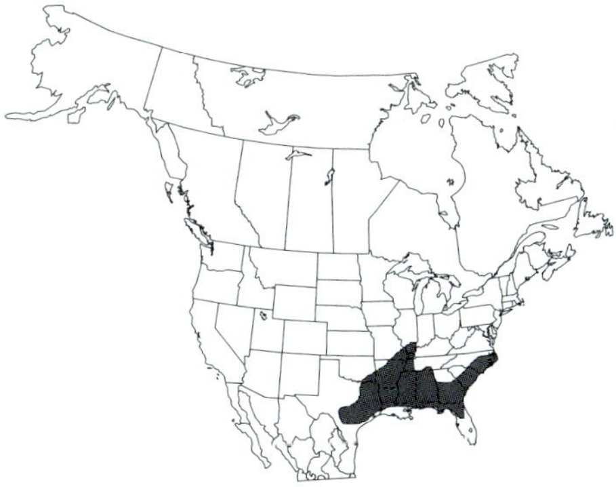

A. pavia

A. pavia

I. Ataxia or Incoordination: horses; apprehension, nervousness, ataxia

Artemisia
sagebrush
Asteraceae, sunflower family

Signs
- horses: apprehension, nervousness, thoracic limb ataxia
- usually transient as signs resolve with continued access to plant
- may also have transient irritant effects on digestive system in horses

Pathology
- no significant lesions

Treatment
- seldom necessary

Disease Problem
- "sage sickness" in horses
- typically useful winter forage for livestock and wildlife
- adverse effects when animals abruptly forced to subsist on or otherwise eat large amounts
- toxins unknown but may be related to array of irritant, aromatic monoterpenes present in plants
- also causes temporary irritant digestive effects in ruminants

Taxonomy and Ecology
- representative species of toxicologic concern:

A. filifolia	sand sage, sand sagebrush
A. ludoviciana	white sage, western mugwort, prairie sage, sagewort
A. spinescens	bud sage, button sage, spring sage
A. tridentata	sagebrush, big sagebrush

- aromatic herbs or shrubs, 30–200 cm tall
- leaves simple, alternate, blades of various shapes
- flowers (florets) borne in numerous, small heads
- ray florets absent or greatly reduced
- disk florets tan or yellowish-white
- variety of soil types in open, dry habitats; may be abundant

Artemisia tridentata

A. tridentata

A. tridentata

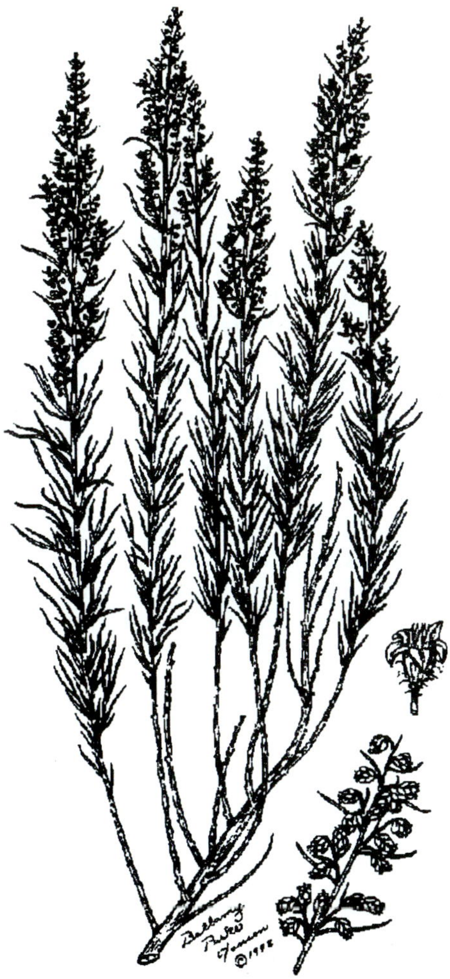

A. filifolia

I. Ataxia or Incoordination: depression; head down as if asleep; stare as if blind; stiff gait; eats slowly and with difficulty; violent response to restraint in horses; loss of sense of direction and flocking instinct in sheep and goats

Astragalus
locoweed
Fabaceae, pea family

Signs and Diagnostic Aids

- depression, stand with head and neck down as if asleep, stare as if blind, gait awkwardly stiff-legged, movement of limbs in exaggerated manner
- eating slow and difficult, appearing to be sucking on plant stems while trying to eat, nibbling with lips quivering, jaws moving spastically, head nodding
- horses: when handled become violent, especially with restraint, shy violently, difficult to back up or lead, rear up and fall backward
- cattle: belligerent but less prone to violent behavior
- weakness, depression, rough coat, weight loss, ataxia, loss of balance, head shaking, sudden movements
- sheep and goats: nervousness, stiffness, pacing gait with head held high, ataxia, loss of flocking instinct
- animals grazing at high elevations may display signs of right heart failure as early signs of locoism
- four- or fivefold increase in serum AST and AP, cytoplasmic vacuolation in peripheral lymphocytes

Pathology

- microscopically: cytoplasmic vacuolation of cells in thyroid, parathyroid, pancreas, adrenals, kidney, liver, lymph nodes, and neurons of central and peripheral nervous systems, especially Purkinje cells of cerebellum

Treatment and Prevention

- violent behavior in horses not controlled by tranquilizers
- 3–4 mg of reserpine i.m. adult horse temporary help
- limit access to plants to early pod stage and stock range heavily so legumes eaten quickly
- allow grazing of infested range at short intervals (cyclic grazing)
- feed affected animals good forage to regain condition

Disease Problem

- "locoism": horses most susceptible
- appears when animals eat 30–50% b.w. in d.w. of plant, usually over several weeks or more
- affected horses should be considered permanently unsafe for riding
- indolizidine alkaloids, swainsonine and its *N*-oxide
- inhibit breakdown of mannose oligosaccharides with accumulation in lysosomes, cell function altered
- also causes selenosis, PEM (blind staggers), reproductive problems

Taxonomy and Ecology

- representative species specifically causing locoism:

A. lentiginosus	speckled loco, freckled milk vetch
A. mollissimus	woolly loco, purple loco
A. wootonii	garboncillo, Wooton's locoweed

- decumbent to erect, annual or perennial herbs
- leaves compound, terminal leaflets and stipules present
- flowers pealike, petals 5, white, pink, violet
- legumes sometimes inflated
- almost exclusively in western half of continent; in grasslands, deserts, alpine tundra, open woods

Astragalus mollissimus

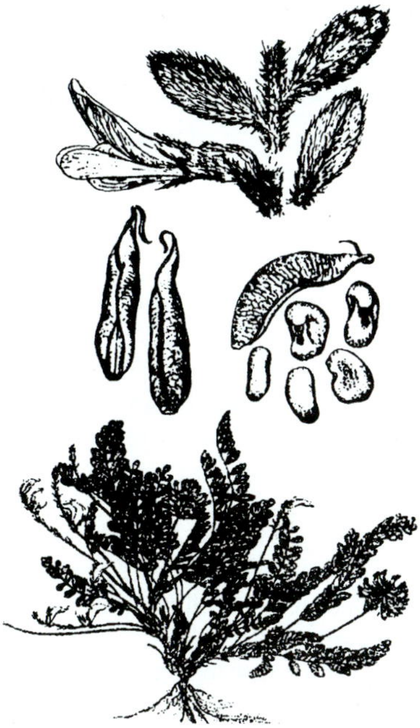

A. mollissimus

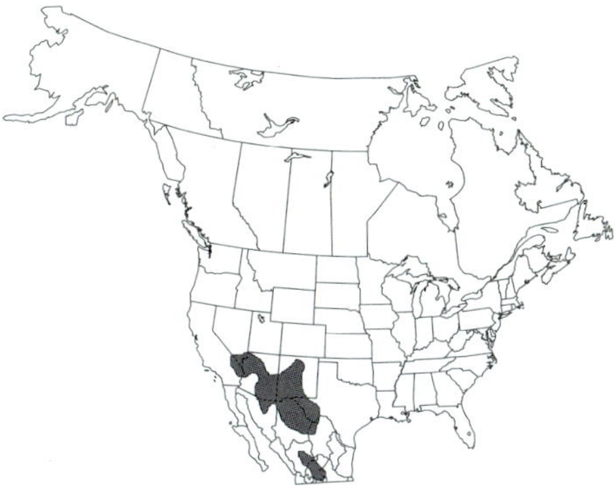

A. wootonii

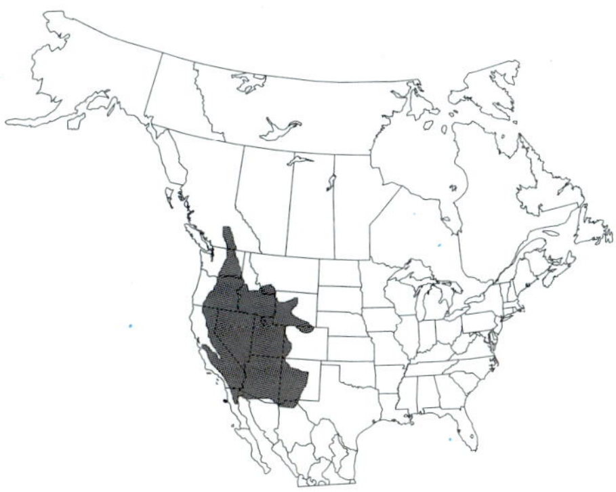

A. lentiginosus

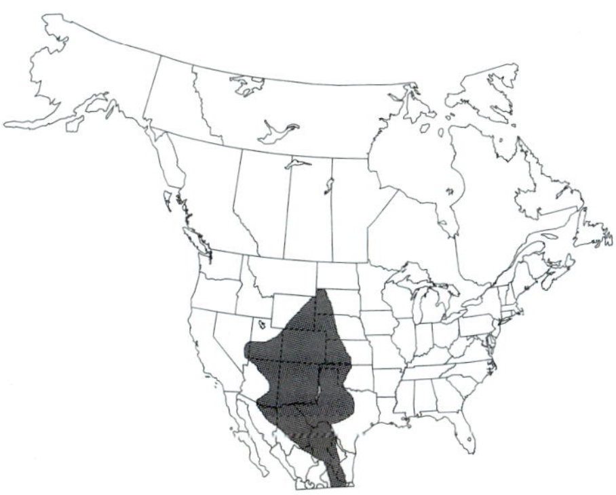

A. mollissimus

I. Ataxia or Incoordination: abrupt onset; ataxia, staggering, recumbency, death; more common gradual onset; weakness, ataxia, pelvic limb placement problem, tremors, dazed appearance, labored respiration, clicking sound when walking

Astragalus
milk vetch
Fabaceae, pea family

Signs and Diagnostic Aids

- abrupt onset; constipation, slight ataxia, staggering walk, temporary blindness, recumbency, and death due to cardiac and respiratory failure
- more common gradual onset; weakness, ataxia, increased salivation, muscle spasms and tremors, dazed and docile appearance, proprioceptive deficit with placement problems especially noticeable in pelvic limbs
- difficult breathing, wheezing, roaring, or coughing, continual flexion of rear fetlocks, distinctive clicking sound made when dew claws or hooves strike opposite rear leg during movement
- when startled, animal may collapse on pelvic limbs or be ataxic
- horses: sweating and paroxysmal episodes of distress, staggering, and falling
- MetHb may be 20% or higher

Pathology

- microscopically: foci of necrosis in brain and degenerative lesions of spinal cord, sciatic nerve

Treatment

- provide nursing care, alternative forage

Disease Problem

- "cracker-heel disease" in forested mountain areas
- cattle or sheep: "timber trouble," "timber grass poisoning," "timber paralysis," "roaring disease," "lungers," "wheezers"
- horses: "mountain fever," "roaring disease," "timber paralysis"
- dosage causing intoxication is much lower than for locoweeds, 1–2% b.w. over several days
- glycosides of 3-nitropropionic acid (3-NPA) and 3-nitropropanol (3-NPOH) (miserotoxin), nitrotoxins
- irreversible inhibition of mitochondrial succinate dehydrogenase, activation of ATP-sensitive potassium channels and depolarization, decreased energy availability in brain

Taxonomy and Ecology

- representative species of toxicologic concern:

A. canadensis	Canada milk vetch
A. emoryanus	Emory's milk vetch
A. miser	timber milk vetch, weedy milk vetch

- decumbent to erect, annual or perennial herbs
- leaves compound, terminal leaflets and stipules present
- flowers pealike, petals 5, white, pink, violet, or yellowish
- legumes sometimes inflated
- seldom abundant, toxic species are almost exclusively in western half of continent; mainly open mountain woodlands

Astragalus miser

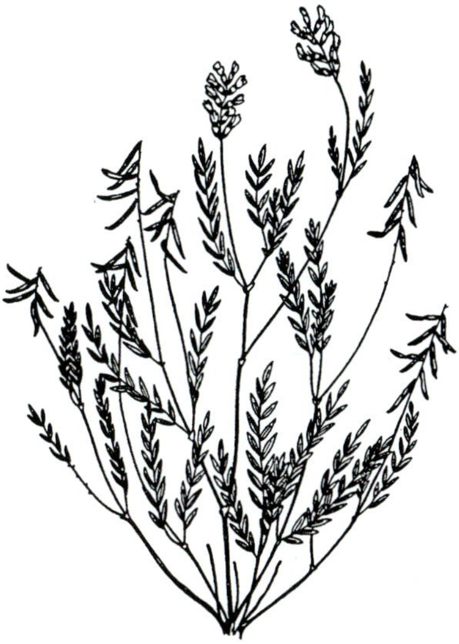

A. miser

A. miser

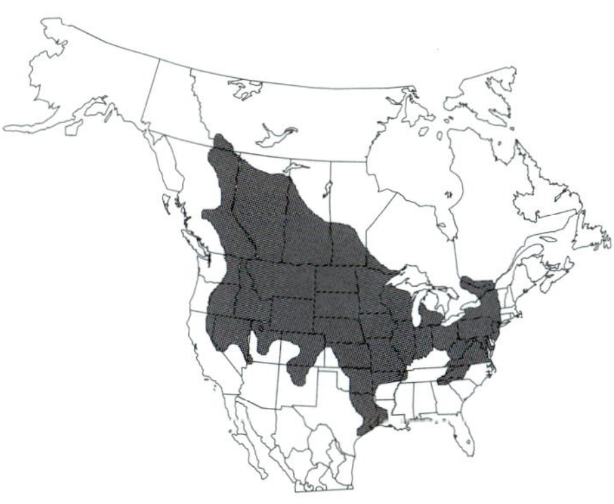

A. canadensis

A. canadensis

I. Ataxia or Incoordination: salivation, colic, loss of appetite, ataxia, knuckling of fetlocks

Laburnum
golden-chain tree
Fabaceae, pea family

Signs
- increased salivation, anorexia, ataxia, tremors, colic
- cattle: knuckling of fetlocks
- dogs: abdominal pain, vomiting without diarrhea, weakness

Pathology
- no distinctive lesions

Treatment
- administer activated charcoal orally
- provide supportive care
- induce emesis in dogs before onset of signs

Disease Problem
- leaves and especially legumes and seeds are toxic
- mild to moderate neurologic dysfunction; toxic effects are not always manifested
- 0.1–1.0% b.w. may cause intoxication
- possibly mild myodegeneration
- quinolizidine alkaloids, cytisine, and/or N-methylcytisine; activate central and peripheral nicotinic cholinergic receptors, stimulation
- teratogenic effects are possible due to the presence of anagyrine

Taxonomy and Ecology
- two species and one hybrid of toxicologic concern:

L. alpinum	Scotch laburnum, alpine golden-chain
L. anagyroides	golden-chain tree, golden-rain tree
L. x *watereri*	golden-chain tree, Voss's laburnum

- deciduous shrubs or small trees
- leaves palmately compound, leaflets 3
- flowers pealike, lemon yellow
- legumes flat with several seeds
- ornamentals; primarily in eastern half of continent

Laburnum anagyroides

L. anagyroides

I. Ataxia or Incoordination: stiff gait, weakness, ataxia, tremors, reluctance to walk

Lathyrus
pea
Fabaceae, pea family

Signs

- vary somewhat among species of *Lathyrus*
- horses: stiff or stringhalt-like gait, ataxia, hyperexcitability, weakness, struggle to stand
- flaccid paraplegia, paresis, or paralysis of larynx with loud inspiratory and expiratory noises
- sheep: abrupt onset of lethargy, anorexia, ataxia, pacing, circling, head pressing, tremors, increased salivation, grinding of teeth, seizures occurring as sudden extension of forelimbs, head, and neck
- cattle: as in sheep, possibly with paraplegia; milder form with *L. hirsutus* when legumes are present; lame, stand hunched up with legs under body, reluctant to walk, prefer to lie down, muscular tremors; transient disease

Pathology

- microscopically: with some plant species; degeneration of motor neurons in brain stem, medulla, and spinal tracts, with focal demyelination, recurrent laryngeal nerve demyelination

Treatment and Prevention

- provide sedation to control seizures
- forage should not be fed when the maturing legumes and seeds present
- if removed from access to plants in early stages, disease usually reversible

Disease Problem

- "neurolathyrism": "osteolathyrism" is mainly an experimental disease
- toxic effects among *Lathyrus* species vary considerably
- in most instances, the plants are good forage when legumes not present
- three neurologic syndromes: acute, sometimes lethal disease with distinctive neuropathologic changes as with *L. latifolius* in horses; acute, often lethal disease without distinct neuropathologic changes as with *L. sylvestris* in sheep; acute, generally nonlethal disease as with *L. hirsutus* in cattle
- novel aminoacids: agonists at central glutamate, kainate, and quisqualate excitatory receptors causing excessive activity leading to neuronal degeneration
- also affects reproductive system

Taxonomy and Ecology

- representative species of toxicologic concern:

L. hirsutus	Caley pea, wild winter pea
L. latifolius	everlasting pea, perennial pea
L. pusillus	singletary pea
L. sativus	chickling pea, riga pea
L. sylvestris	flat pea, wood pea, flat peavine

- annual or perennial, climbing or trailing vines, stems sometimes winged
- leaves 1-pinnately compound, leaflets 2–14, tendrils replacing terminal leaflets, leaflike stipules
- flowers pealike, white, pink, blue, purple, red, or yellow
- legumes usually flattened, 2–many seeds
- most toxic species are cultivated ornamentals, forage, or green manure crops

Lathyrus hirsutus

L. latifolius

L. latifolius

L. palustris

I. Ataxia or Incoordination: depression; head down as if asleep; stare as if blind; stiff gait; eats slowly and with difficulty; violent response to restraint in horses; loss of sense of direction and flocking instinct in sheep and goats

Oxytropis
locoweed
Fabaceae, pea family

Signs and Diagnostic Aids

- depression, stand with head and neck down as if asleep, stare as if blind, gait awkwardly stiff-legged, movement of limbs in exaggerated manner
- eating slow and difficult, appears to be sucking on plant stems while trying to eat, nibbling with lips quivering, jaws moving spastically, head nodding
- horses: when handled, become violent, especially with restraint, shy violently, difficult to back up or lead, rear up and fall backward
- cattle: belligerent but less prone to violent behavior
- weakness, depression, rough coat, weight loss, ataxia, head shaking, sudden movements
- animals grazing at high elevations may display signs of right heart failure as early signs of locoism
- sheep and goats: nervousness; stiffness, pacing gait with the head held high, ataxia, loss of flocking instinct
- four- or fivefold increase in serum AST and alkaline phosphatase, cytoplasmic vacuolation in peripheral lymphocytes, depression of serum α-mannosidase activity

Pathology

- microscopically: cytoplasmic vacuolation of cells in thyroid, parathyroid, pancreas, adrenals, kidney, liver, lymph nodes, cardiac muscle
- degenerated, vacuolated neurons throughout the central and peripheral nervous systems, especially of Purkinje cells of cerebellum

Treatment and Prevention

- violent behavior in horses not controlled by tranquilizers
- 3–4 mg reserpine i.m. to an adult horse, temporary help
- limit access to plants to early pod stage and stock range heavily so legumes eaten quickly
- allow grazing of infested range at intervals (cyclic grazing)
- feed affected animals good forage to regain condition

Disease Problem

- "locoism": horses most susceptible
- appears when animals eat 30–50% b.w. in d.w. of plant, usually over several weeks or more
- under certain circumstances, animals begin to eat locoweeds, once adapted, at greater risk of disease
- indolizidine alkaloids, swainsonine and its *N*-oxide, inhibit breakdown of mannose oligosaccharides with accumulation in lysosomes; cellular function compromised

Taxonomy and Ecology

- species of toxicologic concern:

O. campestris	field crazyweed, yellow locoweed, slender crazyweed
O. lambertii	Lambert's crazyweed, purple locoweed, purple loco
O. sericea	white loco, white point locoweed

- perennial herbs, to 50 cm tall
- leaves 1-pinnately compound, leaflets 9–41
- flowers pealike, petals 5, with sharply beaked keels, white, yellowish, pink, violet, or purple
- legumes cylindrical, lanceolate, or inflated
- occur primarily in arctic, montane, and prairie habitats; greatest abundance in western half of continent

Oxytropis lambertii

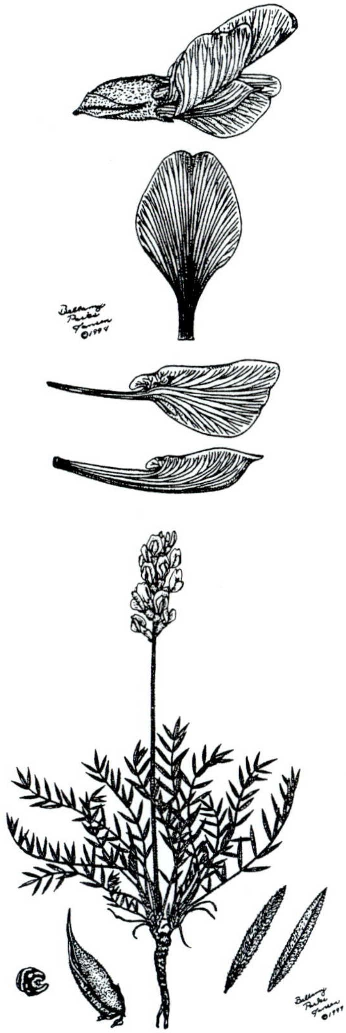

O. lambertii

O. sericea

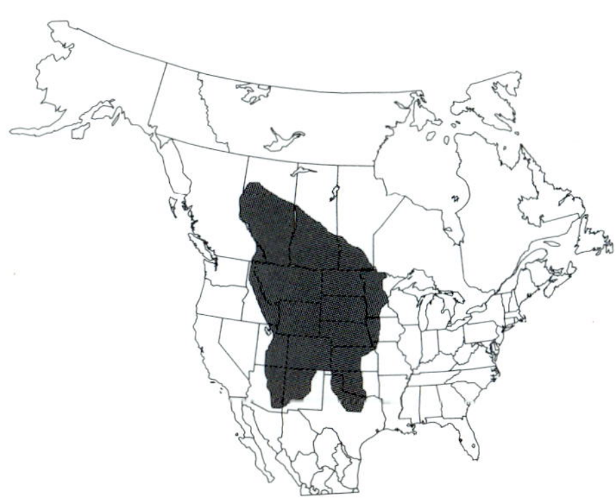

O. lambertii

I. Ataxia or Incoordination: horses; gradual onset, ataxia, twitching, weakness, loss of body condition

Pteridium
bracken fern
Dennstaedtiaceae, bracken family

Signs and Diagnostic Aids
- risk only in horses
- depression, constipation, slightly unsteady gait
- ataxia noticeably worse over next few days, eventual difficulty standing
- marked loss of body condition, weakness, cold extremities, tendency for watery stools, muscular twitching that causes jerky movements of head, neck, limbs, tail
- blood pyruvate and lactate increase severalfold, decrease in blood thiamine and transketolase activity

Pathology
- congestion of brain, viscera with few distinct lesions
- microscopically: scattered neuronal necrosis

Treatment
- administer 0.5–1.0 g thiamine parenterally, for several days

Disease Problem
- fresh or mainly dried plants in hay, 20–30% of diet for 2–3 weeks
- thiamine deficiency; type I thiaminase, base exchange reaction with competitive-type inhibition of thiamine cofactor activity as well as inactivation of thiamine present
- ruminants not affected because of ruminal microbial thiamine synthesis as well as destruction of thiaminases
- also affects hematologic and urinary systems (page 8, 268)

Taxonomy and Ecology
- one species with four varieties of toxicologic concern:

P. aquilinum	bracken fern, brake fern, bracken, braken, lacy bracken, tailed bracken

- deciduous ferns, forming large, dense colonies
- rhizomes woody, black
- fronds erect, pinnately compound, 0.5–1.5 m long
- sporangia partially concealed by revolute margins
- well-drained soils of open woodlands across continent

Pteridium aquilinum

P. aquilinum

P. aquilinum

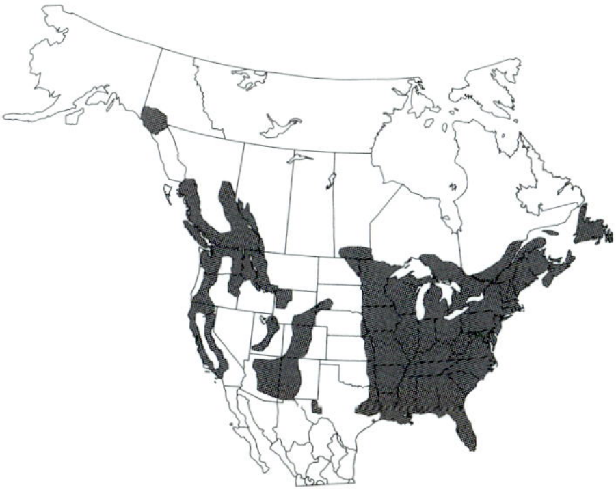

P. aquilinum

II. Depression or Somnolence: disorientation, drowsiness, intermittent periods of arousal, hyperactivity, biting, shivering

Cannabis
marijuana
Cannabaceae, hemp family

Signs
- depression, drowsiness
- interspersed periods of arousal, bizarre actions including hyperactivity, tremors, hypersensitivity to touch and sound
- appear glassy-eyed, dizzy, or disoriented
- compulsive eating, bumping into furniture, biting objects, shivering, decreased body temperature, rigid sawhorse stance
- cats: increased likelihood of neurologic stimulation
- recovery in a few to 24 hours

Pathology
- no lesions

Treatment
- typically no treatment needed
- administer activated charcoal orally, bowel evacuation and/or purgatives
- maintain body temperature

Disease Problem
- inadvertent or intentional exposure of pets
- most animal species susceptible
- impairment of reaction time, motor coordination, and visual perception
- effects on respiration, endocrine function, and the activity of other systems
- Δ-9- and Δ-8-tetrahydrocannabinol (THC)
- low toxicity, rapid metabolism and elimination

Taxonomy and Ecology
- one polymorphic species of toxicologic concern:

C. sativa	hemp, Indian hemp, marijuana, cannabis, kif, hashish, dagga, ganja, pot

- odoriferous, erect herb, 1–5 m tall
- leaves palmately compound, leaflets 5–11, alternate above, opposite below
- illegal clandestine cultivation; naturalized in disturbed sites of upper Midwest

Cannabis sativa

C. sativa

II. Depression or Somnolence: weakness, drowsiness, depression, ventral edema, loss of appetite

Solanum
nightshade
Solanaceae, nightshade family

Signs
- mainly cattle: depression, loss of appetite, drowsiness, ataxia, weakness, labored respiration, abdominal pain
- calves: edema, ventral swelling, excess abdominal fluid, perhaps diarrhea ± blood
- sometimes mild cholinergic effects

Pathology
- reddening, necrosis of mucosa of stomach, intestines
- fluid accumulation in body cavities, abdomen
- rumen flaccid, distended with nightshade berries
- microscopically: Purkinje cell loss in "crazy cow syndrome"

Treatment
- provide nursing care to alleviate signs

Disease Problem
- possible toxic interactions with ivermectin in horses
- all plant parts but immature berries usually most toxic
- most animal species are susceptible but sheep and goats are less likely to be affected
- 27-C pseudoalkaloids or glycoalkaloids (GAs), two major groups: spirosolanes (solasodine, tomatidine) and solanidanes (solanidine)
- potatoes; solanidines; α-chaconine, α-solanine; highest levels fruits, sprouts, foliage at time of flowering
- "crazy cow syndrome": *S. dimidiatum*; calystegins, potent enzyme inhibitors; following unexpected stimulus; tremors, ataxia, collapse, nystagmus, struggle to stand; able to rise after a brief rest
- also affects digestive system and eyes
- similar depression effects may occur with consumption of *Lycopersicon*, cultivated tomato vines

Taxonomy and Ecology
- representative species of toxicologic concern:

S. americanum	black nightshade, common nightshade, stubbleberry
S. carolinense	horsenettle, bullnettle, Carolina horsenettle
S. dimidiatum	western horsenettle
S. elaeagnifolium	silverleaf nightshade, white horsenettle
S. nigrum	black nightshade, poisonberry
S. ptycanthum	black nightshade,
S. rostratum	buffalobur, Texas thistle
S. tuberosum	Irish potato, white potato

- herbs or shrubs, to 1 m tall
- leaves alternate or paired
- flowers showy, corollas 5-lobed, yellow, white, blue, or purple, stamens 5, clustered, opening by apical pores
- fruits globose berries
- widespread across continent; variety of habitats, especially abundant in southern Great Plains

Solanum elaeagnifolium

S. carolinense

S. ptychanthum

S. elaeagnifolium

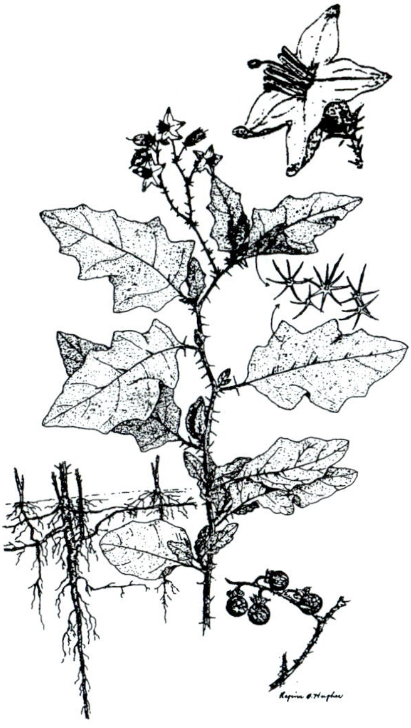

S. carolinense

V. Limb Abnormalities: pelvic limbs of horses; abrupt onset of involuntary flexion and delayed extension of the hocks, bilateral, especially evident when backing or turning

Hypochaeris
cat's ear
Asteraceae, sunflower family

Signs
- horses only: abrupt onset of sudden, involuntary flexion and delayed extension of the hocks, usually bilateral, especially noticeable when animal is backing or turning
- peculiar bunny-hop gait
- sometimes laryngeal hemiplegia

Pathology
- microscopically: Wallerian degeneration of large diameter myelinated fibers of peripheral nerves, peroneal, sciatic, and recurrent laryngeal
- lower spinal axonal degeneration, hypertrophy of long and lateral digital extensors and lateral deep digital flexor muscles

Treatment
- spontaneous recovery may occur weeks to years later
- tenotomy of lateral digital extensors of rear leg, only modifies expression of disease
- phenytoin (10–15 mg/kg b.w.) orally mixed with feed, once or twice daily for two weeks is of some value

Disease Problem
- "reversible or Australian stringhalt"
- *Hypochaeris* only tenuously linked as cause of disease, not confirmed experimentally
- cause of disease unknown

Taxonomy and Ecology
- one introduced species of toxicologic concern:

H. radicata	spotted cat's ear, hairy cat's ear, gosmore, flatweed, rough cat's ear

- herb with milky, viscid sap, to 60 cm tall
- leaves simple, alternate, in basal rosettes
- flowers (florets) borne in showy heads on long stalks
- ligulate florets bright yellow
- naturalized invader of disturbed sites, roadsides, lawns, and fields in Pacific states mainly west of Sierra Nevada/Cascade range

Hypochaeris radicata

H. radicata

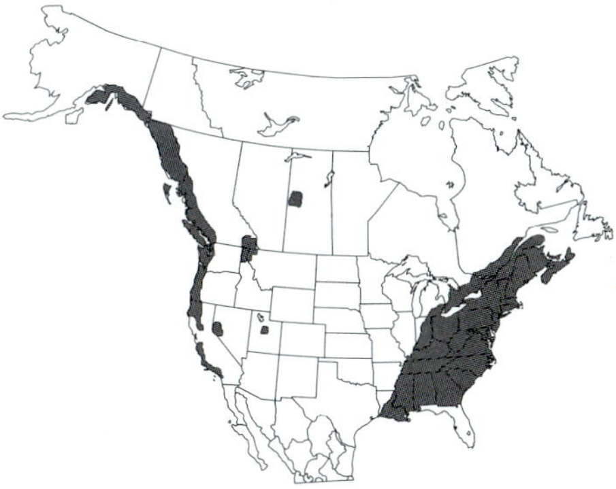

H. radicata

V. Limb Abnormalities: pelvic limbs of ruminants; persistent knuckling-over of fetlocks, weakness, ataxia; continue to eat and drink

Kallstroemia
caltrop
Zygophyllaceae, caltrop family

Signs
- knuckling-over of fetlocks of pelvic limbs in cattle, sheep, and goats; some cases persisting for months
- sheep and goats: may stand and walk on carpal joints
- weakness, pelvic limb ataxia, flexure of hocks, eventually paralysis
- animals unable to stand, continue to eat and drink
- signs may persist for weeks or months
- terminal seizures

Pathology
- no distinctive lesions

Treatment
- provide supportive nursing care to maintain animals until recovery

Disease Problem
- mainly in late summer when plants are major portion of diet for 1 or more weeks
- erratic in occurrence
- toxins unknown, possibly indole alkaloids

Taxonomy and Ecology
- two species of toxicologic concern:

K. hirsutissima	hairy caltrop, carpetweed
K. parviflora	warty caltrop

- prostrate or decumbent herbs, from stout taproot, diffusely branched, to 1 m long and 20 cm tall
- leaves simple, opposite, one of each pair smaller or absent
- flowers solitary, sepals 5, petals 5, white to bright orange
- fruits long-beaked schizocarps
- disturbed soils; most common in southwestern part of continent

Kallstroemia parviflora

K. hirsutissima

K. hirsutissima

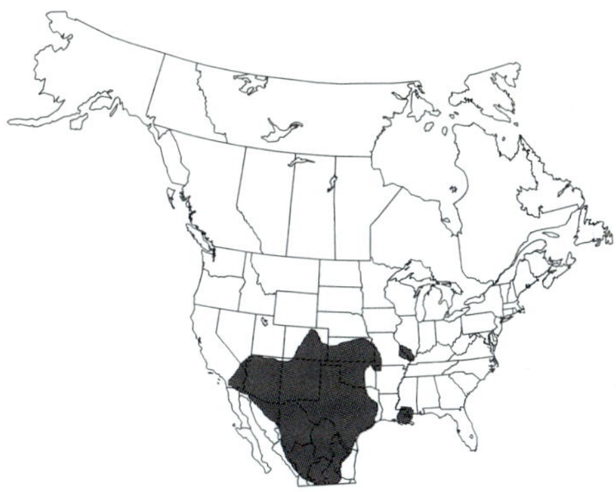

K. parviflora

VI. Limb Abnormalities: paresis/paralysis/muscle weakness, primarily cattle; ataxia, tremors, episodic cycle of falling, struggling, resting, rising, and falling again

Delphinium
larkspur
Ranunculaceae, crowfoot family

Signs

- episodic weakness; animal falls, struggles, rests, rises, and falls again
- signs appear 3–4 hours after ingestion, fully developed in 5–8 hours
- abrupt onset of uneasiness; wide, straddle-legged stance, gait stiff and staggering, repeated falling with kicking, tremors, ataxia, constipation, recumbency, and bloat especially when animal is lying facing downhill
- six stages of intoxication: 1—slight tremors; 2—cannot stand for long periods so periodically lies down; 3—can lift body but cannot stand; 4—remains sternal; 5—is in lateral recumbency and unable to rise to sternal; 6—death
- respiratory difficulties are conspicuous and death is due to respiratory failure complicated by bloat

Pathology

- no distinctive lesions
- presence of plant fragments in rumen
- assay ruminal fluid for alkaloids

Treatment

- administer atropine and physostigmine 0.04–0.08 mg/kg b.w. i.p., repeat as needed; or administer 65 mg physostigmine salicylate, 130 mg pilocarpine, and 32 mg strychnine per 250 kg b.w. s.q.
- animals that fall should be turned to face uphill
- resistance of sheep provides protection, eat early growth

Disease Problem

- mainly cattle; sheep and horses at much lower risk
- risk mainly during flowering and early fruit formation when plants both toxic and palatable
- less than 0.5% b.w. or 20% of diet for a short period may cause intoxication
- diterpenoid alkaloids; methyllycaconitine (MLA), nudicauline; curarelike competitive neuromuscular blocking effects, inhibit acetylcholine at nicotinic postsynaptic receptor sites with loss of motor function

Taxonomy and Ecology

- representative species of toxicologic concern:

D. andersonii	desert larkspur, Anderson's larkspur
D. barbeyi	tall larkspur, duncecap larkspur, western larkspur
D. carolinianum	plains larkspur, prairie larkspur
D. geyeri	plains larkspur, Geyer's larkspur
D. glaucum	mountain larkspur, pale larkspur, western larkspur
D. nuttallianum	meadow larkspur, blue larkspur, Nelson larkspur, dwarf larkspur

- erect, perennial herbs, 50–200 cm tall
- leaves simple, alternate, palmately divided
- flowers in terminal racemes or panicles; sepals 5, 1 spurred; petals 4, upper 2 spurred, various colors
- fruits curved, cylindrical follicles
- distributed across continent; greatest diversity and abundance in western mountain ranges; some taxa forming large stands and becoming locally abundant

Delphinium carolinianum

D. barbeyi

D. geyeri

D. glaucum

VI. Limb Abnormalities: paresis/paralysis/muscle weakness; gradual onset of limb weakness, loss of motor control of limbs; stumbling/stiff gait, hypersensitivity, recumbency

Karwinskia
coyotillo
Rhamnaceae, buckthorn family

Signs
- gradual onset, days to weeks
- increased alertness, hypersensitivity to touch and sounds, tremors, weakness of limbs; stumbling, staggering gait; dragging of pelvic limbs, loss of motor control, rapid progression to flaccid paralysis, collapse
- sensory and autonomic functions normal; eating, drinking, rumination, urination, defecation not affected

Pathology
- microscopically: Wallerian degeneration, demyelination, swelling and degeneration of Schwann cells of large motor nerves, swollen spinal axons of white and gray areas

Treatment
- administer parenteral thiamine hydrochloride daily for possible benefit
- provide nursing care to allow nerve regeneration and/or compensatory adjustment

Disease Problem
- all animal species affected; often children eating seeds with fruit
- intoxication 1–2 weeks after initial ingestion
- all plant parts toxic, leaves least, seeds most
- effects cumulative, 0.1% b.w. of seeds may cause serious disease
- 1% b.w. leaves will cause less pronounced signs
- four or more complex anthracenones, T-496, T-514, T-516, T-544 (tullidinol)
- "dying back syndrome" due to disruption of axoplasmal transport, degeneration of distal axonal segment, myelin degeneration, of large peripheral motor nerves
- progressive polyneuropathy with ascending systemic paralysis and flaccid quadriplegia; death due to respiratory arrest

Taxonomy and Ecology
- two species of toxicologic concern:

K. humboldtiana	tullidora, coyotillo, buckthorn
K. parviflora	coyotillo

- erect shrubs or small trees, 1–3 m tall
- leaves simple or opposite, with small translucent glands
- flowers perfect, sepals 5, petals 5, clawed
- fruits purple to black drupes, 2–3 stones
- plains and brushy areas of southwestern Texas into Mexico

Karwinskia humboldtiana

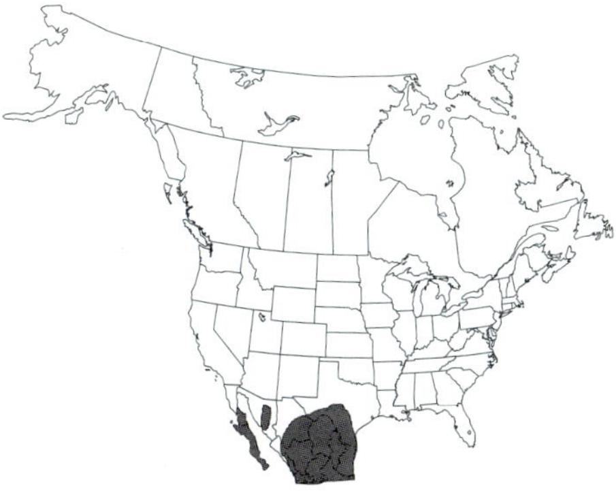

K. humboldtiana

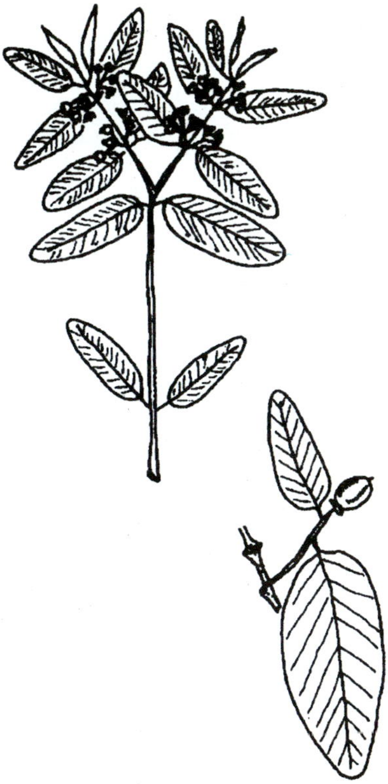

K. humboldtiana

VI. Limb Abnormalities: paresis/paralysis/muscle weakness, recumbency, depression, tremors

Thermopsis
golden banner
Fabaceae, pea family

Signs and Diagnostic Aids

- depression, anorexia, arched back, tremors, rough haircoat, drawn-in flanks, swollen eyelids
- persistent recumbency for several days to week or more, continue to eat and drink if food and water are available
- increase in serum CK, LDH, AST
- death likely from lack of water

Pathology

- pale streaks in muscles
- microscopically: hyaline degeneration and necrosis of skeletal muscle

Treatment

- administer activated charcoal orally
- disease reversible, provide feed and water

Disease Problem

- plants grazed for several days or more, 0.1–0.3% b.w.
- toxicity varies with growth stage of plant; flowers and legumes greatest risk
- α-pyridone quinolizidine alkaloids; anagyrine, thermopsine, 5,6-dehydrolupanine, cytisine, N-methylcytisine, lupanine, and 17-oxosparteine
- high concentrations of anagyrine may cause teratogenic effects similar to those caused by *Lupinus*

Taxonomy and Ecology

- one polymorphic species of toxicologic concern:

T. rhombifolia	round-leaved thermopsis, prairie buck bean, yellow pea, false lupine, golden banner

- erect perennial herb, to 1 m tall
- leaves palmately compound, leaflets 3, stipules large
- flowers showy, pealike, petals 5, yellow or purple
- mountain meadows, desert shrub communities, and grasslands; Rocky Mountains to Pacific Coast

Thermopsis rhombifolia

T. rhombifolia

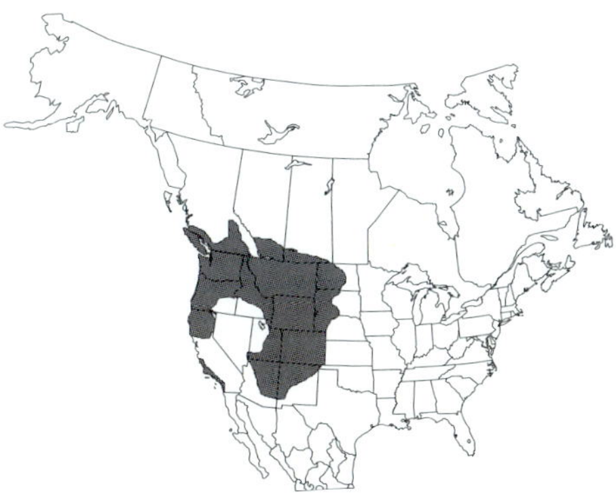

T. rhombifolia

VIII. Lip and Tongue Movement Abnormalities: difficulty eating, drinking, and swallowing; horses only; abrupt onset, hypertonic lips and tongue, constant chewing

Centaurea
knapweed
Asteraceae, sunflower family

Signs
- abrupt onset of impaired eating, drinking, apprehension
- pronounced depression, unilateral or bilateral hypertonicity of lips, tongue; fixed facial expression, continuous chewing
- mouth open or tightly closed, tongue protruding curled to form a groove or V shape
- locomotor difficulties, aimless walking, drowsy, head low
- disease permanent, but not necessarily progressive after the first few days
- dehydration, starvation

Pathology
- globus pallidus, substantia nigra: pale yellowish to buff-colored foci of softening and/or cavitation, to 1 cm in diameter, typically bilateral and symmetrical
- microscopically: extensive necrosis of neurons, glia, and capillaries within sharply defined margins

Treatment
- provide general nursing care
- provide good feed and water, vitamin supplements

Disease Problem
- "chewing disease" or "equine parkinsonism"
- appears after continuous, long-term ingestion of exceptionally large amounts of plant, 60–100% b.w.
- suspect guaianolide sesquiterpene lactones; repin, subluteolide, janerin, cynaropicrin and acroptilin; highly reactive epoxides, unstable precursors
- also affects digestive system

Taxonomy and Ecology
- three species of toxicologic concern:

C. melitensis	Maltese star thistle
C. repens	Russian knapweed, Turkistan thistle, Russian star thistle
C. solstitialis	yellow star thistle, St. Barneby's thistle

- annual or biennial herbs, 30–175 cm tall
- leaves simple, alternate, upper usually reduced
- flowers (florets) borne in numerous heads at branch ends
- ray florets absent
- disk florets yellow or purple
- aggressive invaders of pasture and disturbed areas; may form large populations mainly in western half of continent

Centaurea solstitialis

C. repens

C. solstitialis

C. repens

VIII. Lip and Tongue Movement Abnormalities: difficulty eating, drinking, and swallowing; gradual onset of weight loss, chewing difficulty with tongue protruding from mouth

Prosopis
mesquite
Fabaceae, pea family

Signs

- gradual onset; conspicuous weight loss, hair loss, anemia, loss of appetite
- nervous, difficulty coordinating jaw and tongue movements, tongue often protruding several cm from mouth
- continuous chewing, regurgitation of beans
- rumen inactive, profuse stringy salivation
- animals may recover but subject to relapse

Pathology

- wasting, starvation-type lesions especially masseters
- rumen impacted with pods, seeds
- omasum and abomasum nearly empty

Treatment

- administer oral glucose and electrolyte solutions
- provide rumen inoculation using ingesta from healthy animals

Disease Problem

- pods eaten without problem, except when livestock subsist on them for months
- goats especially susceptible and may survive for only 2–3 weeks
- disease typically reversible in cattle
- sheep low risk
- toxicant unknown, possibly tyramines
- also affects digestive system and skin

Taxonomy and Ecology

- one species of toxicologic concern:

P. glandulosa	honey mesquite, mesquite

- shrub or small tree, stout spines at nodes
- leaves alternate, 2-pinnately compound
- flowers fragrant, creamy white to pale yellow
- legumes long, cylindrical, constricted between seeds
- deep rooted, drought and heat tolerant; typically in arid and semiarid areas; washes, rocky ridges, and grasslands of southwestern part of continent

Prosopis glandulosa

P. glandulosa

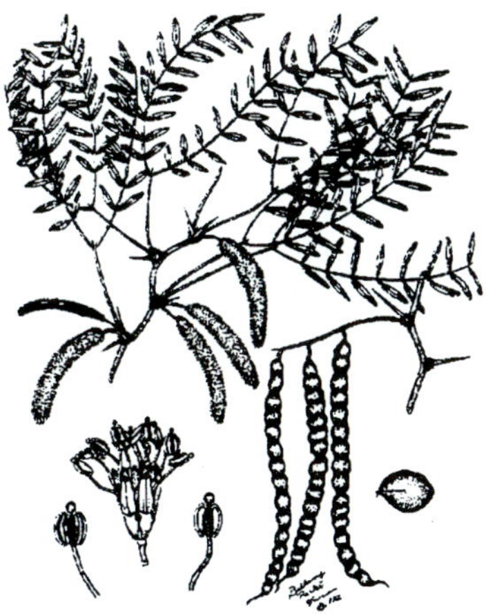

P. glandulosa

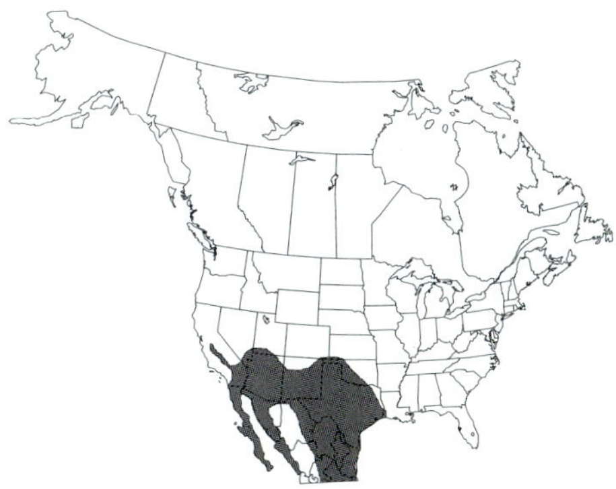

P. glandulosa

VIII. Lip and Tongue Movement Abnormalities: difficulty eating, drinking, and swallowing; abrupt onset of apparent blindness

Zanthoxylum
prickly ash
Rutaceae, rue family

Signs
- apparent blindness, high-stepping gait
- inability to drink and swallow
- eventual recumbency with struggling
- some animals die in a few days, most recover in a few weeks

Pathology
- no distinctive lesions

Treatment
- provide supportive nursing care
- calcium preparations and acetylcholinesterase inhibitors, such as neostigmine, may provide some benefit

Disease Problem
- bark has irritant and neurotoxic effects; leaves also possibly toxic
- bark: local anesthetic properties when chewed
- toxin unknown; possibly neuromuscular/ganglionic blocker similar in action to quaternary ammonium compounds
- also contains alkaloid nitidine, inhibitor of brain NaK-ATPase
- also affects eyes

Taxonomy and Ecology
- representative species of toxicologic concern:

Z. americanum	toothache tree, northern prickly ash
Z. clava-herculis	Hercules' club, southern prickly ash, toothache tree
Z. flavum	yellowheart, satinwood, yellowheart prickly ash
Z. fagara	wild lime, colima, una de gato

- shrubs or trees; deciduous or evergreen, aromatic, 1–17 m tall
- leaves pinnately compound, alternate, glandular-punctate
- flowers in small paniculate cymes
- petals 4–5 or none, yellow-green
- fruits follicles
- moist woods and thickets; eastern half of continent, or in Florida Keys and along Gulf Coast

Zanthoxylum americanum

Z. americanum

Z. americanum

Z. americanum

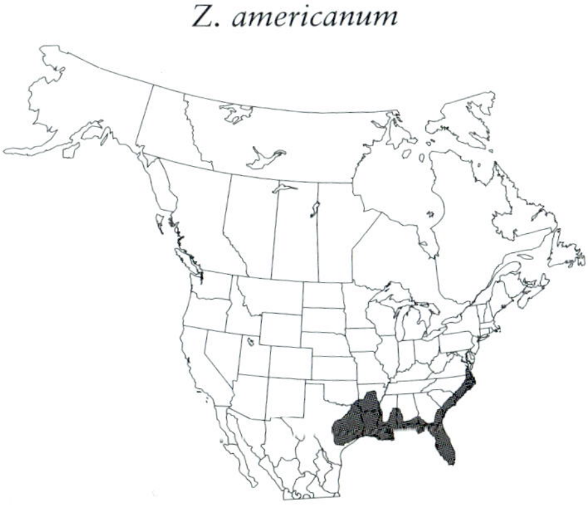

Z. clava-herculis

IX. Pacing, Head Pressing, Circling, Chewing, Yawning: polioencephalomalacia (PEM); ruminants; anorexia, ataxia, blindness, ear twitch and droop

Kochia
kochia
Chenopodiaceae, goosefoot family

Signs and Diagnostic Aids

- abrupt onset; twitching of ears and muscles of the face, champing of the jaws with profuse salivation; animals stand motionless in apparent stupor, circle, or wander aimlessly forward or backward, appear blind; aggressive/excitable behavior, collapse, seizures
- unable to eat properly
- increase in serum pyruvic and lactic acid
- low morbidity, high case fatality rate

Pathology

- herniation of cerebellum into foramen magnum, swollen, flattened gyri, softness and liquifaction of the cerebral cortex gray but not white matter
- microscopically: edema, neuronal necrosis and gliosis, degeneration of granular layers and Purkinje cells
- brain fluoresces under ultraviolet light, a nonspecific test for cerebrocortical necrosis

Treatment

- partial response to parenteral thiamine HCl, 5–10 mg/kg b.w. or 1 g/sheep or more in cattle

Disease Problems

- associated mainly with green forage eaten for several weeks
- sporadic occurrence and in only a few animals
- possibly due to complex interaction between plant and soil sulfate
- mechanism of toxicity unknown, possibly lipid peroxidation
- additional species associated with PEM: *Astragalus* (page 128), *Brassica* (page 14), *Conyza*, *Phalaris* (page 200)
- also affects digestive system and skin (photosensitization)

Taxonomy and Ecology

- one introduced species of toxicologic concern:

K. scoparia	burning bush, summer cypress, fireball, fireweed, Mexican fireweed, kochia

- herbaceous annual from taproot, to 2 m tall
- stems profusely branched, purplish red at maturity
- leaves simple, alternate, narrow
- flowers small, in axillary leaflike bracts at branch ends
- common in disturbed soils; naturalized across western half of continent; cultivated as a forage

Kochia scoparia

K. scoparia

K. scoparia

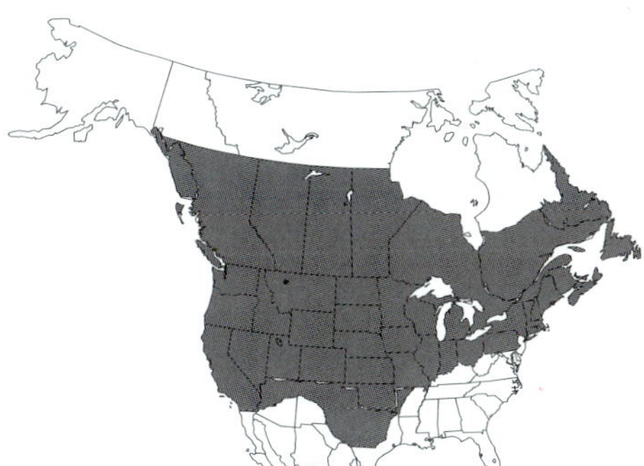

K. scoparia

X. Seizures: abrupt onset; hypersensitive to stimuli, twitching, tremors, periodic seizures with collapse and tonic and clonic movement

Albizia
mimosa tree
Fabaceae, pea family

Signs
- exaggerated response to tactile, auditory, and visual stimuli; muscular twitching and tremors
- seizures characterized by deep, labored respiration, apparent excess salivation or foaminess due to jaw movements, backing up or turning, bizarre neck postures, collapse with extended thoracic limbs and paddling pelvic limbs
- seizures may last 1–2 min and occur again in several minutes

Pathology
- no distinctive lesions

Treatment
- administer parenteral pyridoxine HCl (10–20 mg/kg b.w.); second dose in 8–12 hours

Disease Problem
- intoxication from 1–2% b.w. of dry, mature legumes
- risk with leaves unknown
- cattle, sheep, and goats
- risk for other species unknown although probable risk for dogs
- signs 12–24 hours after ingestion
- alkaloids; structural analogues of vitamin B_6; possibly interfere with GABA synthesis

Taxonomy and Ecology
- one introduced species of toxicologic concern:

A. julibrissin	silk tree, mimosa tree, pink siris

- small spreading tree, to 9 m tall
- leaves 2-pinnately compound, no terminal leaflets
- flowers pink or white, petals 4 or 5
- legumes thin walled, flat, seeds cause bulges in walls
- cultivated ornamental yard tree; naturalized in southeastern United States

Albizia julibrissin

A. julibrissin

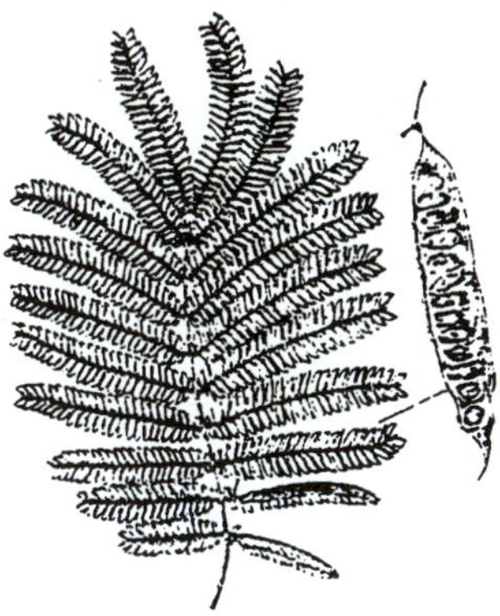

A. julibrissin

A. julibrissin

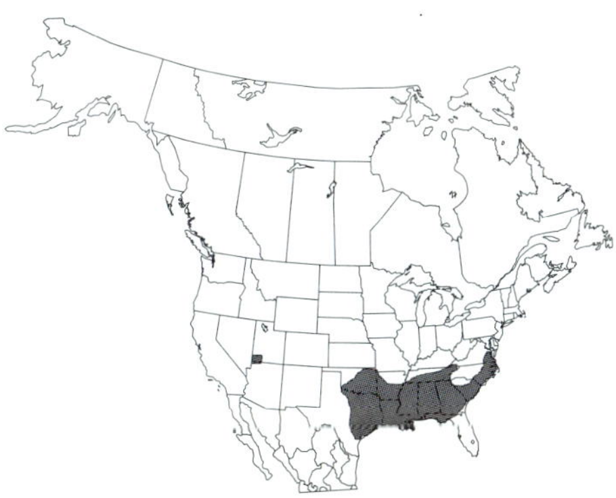

A. julibrissin

X. Seizures: abrupt onset; horses and cattle: weakness, uneasiness, ataxia, tremble, tetanic or clonic leg movement; sheep: humped up, champing jaws, head shaking; twitching of ears, lips, and eyelids

Asclepias
whorled-leaved milkweeds
Asclepiadaceae, milkweed family

Signs
- horses: colic, uneasiness, ataxia, weakness of pelvic limbs
- trembling, falls, mydriasis, profuse sweating
- agonizing tetanic seizures, neck bowed and thoracic limbs flexed, later clonic types with paddling movements
- cattle: less-severe neurologic signs
- sheep: stand with humped-up appearance, head elevated, opisthotonus, champing of jaws; twitching of eyelids, ears, and lips; head shaking, head pressing
- poultry: loss of muscular control, excitement, bizarre posturing, periodic violent convulsions

Pathology
- no distinctive lesions except traumatic skin damage

Treatment
- provide sedation

Disease Problem
- caused by whorled-leaved taxa, especially *A. fascicularis* and *A. subverticillata*
- animals may eat fresh; more readily accept in hay
- toxicity not appreciably reduced by drying
- all animal species, 0.2–1% b.w., horse especially at risk
- identity of toxin not confirmed, possibly pregnane glycosides

Taxonomy and Ecology
- five species of toxicologic concern:

A. fascicularis	narrow-leaved milkweed, Mexican milkweed
A. mexicana	Mexican whorled milkweed
A. pumila	plains whorled milkweed
A. subverticillata	western whorled milkweed, horsetail milkweed
A. verticillata	eastern whorled milkweed, spider milkweed

- erect, perennial herbs with white latex-type sap, to 1.5 m tall
- leaves whorled, mostly narrow
- flowers in umbellate clusters
- petals of various colors, spreading or reflexed
- coronas hood-shaped, erect or incurved
- fruit fusiform follicles
- wide variety of habitats across continent, greatest abundance southward

Asclepias subverticillata

A. subverticillata

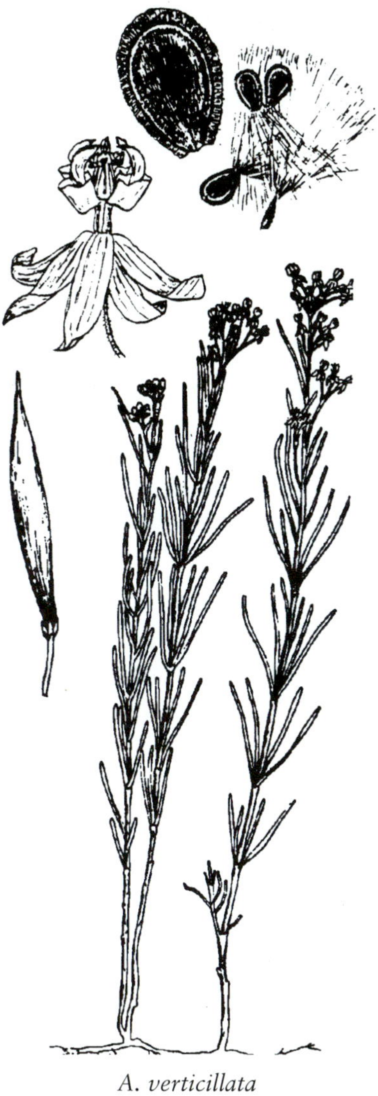

A. verticillata

A. subverticillata

X. Seizures: abrupt onset; violent trembling of entire body following forced movement

Astrolepis
jimmy fern
Pteridaceae, maidenhair fern family

Signs
- no signs while quietly grazing; but when forced to move for a few minutes, abrupt onset of violent trembling of entire body
- back first arches, then a few stilted steps backward, and seizure begins
- sudden death due to respiratory failure, or signs subside after a few minutes rest only to reappear when animal is forced to move again
- 1–2 weeks for recovery from tendency for seizures to occur

Pathology
- no distinctive lesions

Treatment
- provide supportive nursing care
- move animals carefully and avoid undue stress of exercise

Disease Problem
- "jimmies"; primarily sheep, occasionally goats and cattle
- primarily due to ingestion of *A. cochisensis*
- less than 1% b.w. over 2–3 days, fresh or dry, is toxic
- because it is evergreen and somewhat palatable, intoxications occur typically in winter
- morbidity may be high, case fatality rate usually low, especially if avoid undue animal stress
- toxin unknown; appears to be eliminated in milk

Taxonomy and Ecology
- three species of toxicologic concern:

A. cochisensis	jimmy fern
A. integerrima	cloak fern
A. sinuata	wavy cloak fern, long cloak fern, bulb cloak fern

- evergreen herbs from short rhizomes
- fronds 1-pinnately compound, erect, leathery, waxy, densely clustered, 7–130 cm long
- sori separate; submarginal
- slopes and summits of limestone hills and mountains

Astrolepis cochisensis

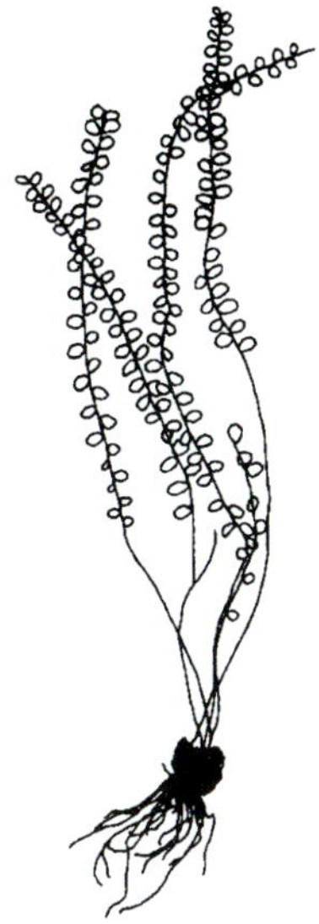

A. cochisensis

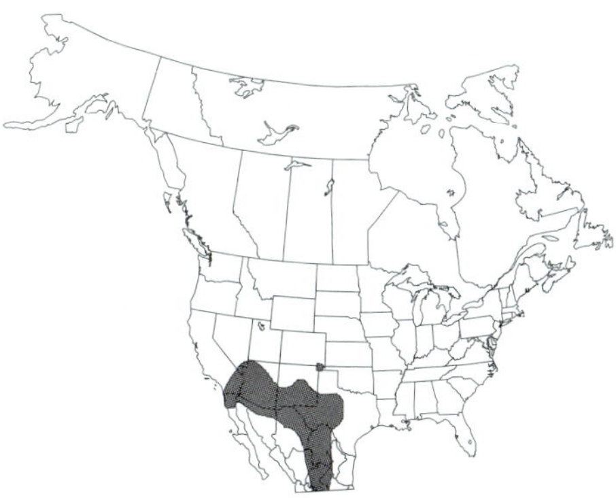

A. cochisensis

X. Seizures: abrupt onset; weakness, loss of appetite, diarrhea, stiffness, trembles, staggers, seizures

Baccharis
baccharis
Asteraceae, sunflower family

Signs
- weakness, ruminal stasis, loss of appetite, excessive salivation, diarrhea
- animals walk as if sore and stiff, prefer to lie down
- when forced to move, tremble, stagger, and may have seizures

Pathology
- reddening, edema, possibly ulceration of abomasum

Treatment
- seldom required
- provide relief of diarrhea

Disease Problem
- mainly cattle forced to subsist on plants in autumn and early winter or when other forage not available
- other ruminants also at risk
- 1% b.w. may cause intoxication
- specific toxin unknown
- also affects digestive system

Taxonomy and Ecology
- representative species of toxicologic concern:

B. angustifolia	baccharis, false-willow, silverling, narrowleaf baccharis
B. glomeruliflora	baccharis, silverling, southern baccharis, groundsel-bush
B. halimifolia	eastern baccharis, groundsel tree
B. pteronioides	yerba manza, yerba del pasmo

- perennial subshrubs or shrubs, 0.5–4 m tall
- leaves simple, alternate
- flowers (florets) borne in numerous heads
- disk florets only; white, pink, or yellowish brown
- commonly associated with wet sites—marshes, washes, streambeds, streambanks, and ditches

Baccharis halimifolia

B. halimifolia

B. pteronioides

B. pteronioides

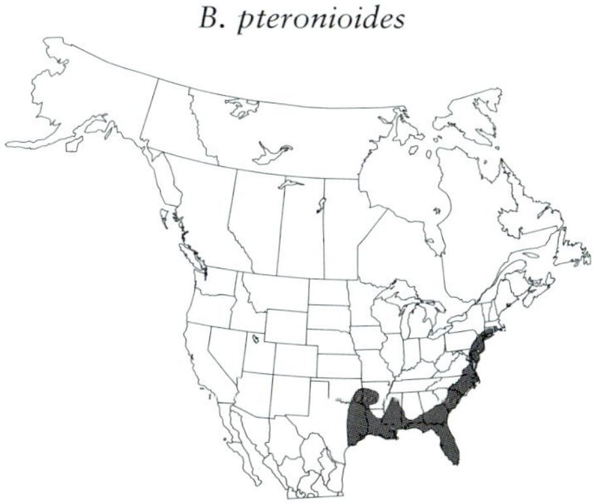

B. halimifolia

X. Seizures: abrupt onset; anxiety, excitement, coughing, gagging, tetanic and paddling seizures

Brunfelsia
yesterday-today-tomorrow
Solanaceae, nightshade family

Signs

- anxiety and excitement followed by persistent coughing, sneezing, and gagging, less often nystagmus
- after a few hours, tremors begin, followed by repeated tetanic-type seizures and occasional paddling seizures
- seizures precipitated by external stimuli similar to those caused by strychnine
- period of tremors and seizures may last hours or days
- complete recovery may take a week or more

Pathology

- no specific lesions

Treatment

- mainly symptomatic; sedation with pentobarbital or diazepam to control seizures
- administer fluids, electrolytes orally
- administer activated charcoal orally

Disease Problem

- probably all animal species are affected; dogs relish fruits and appear to be at greatest risk
- neuroexitant brunfelsamidine likely toxin; scopoletin and neurodepressant also present

Taxonomy and Ecology

- representative species of toxicologic concern:

B. americana	lady-of-the-night, Franciscan rain tree
B. australis	yesterday-today-tomorrow, morning-noon-and-night, Paraguayan jasmine
B. pauciflora	yesterday-today-tomorrow, morning-noon-and-night, yesterday and today

- winter-flowering evergreen shrubs or small trees
- leaves simple, alternate, thick, leathery
- flowers showy, solitary or clustered at branch ends
- corollas 5-lobed, violet, purple, or blue changing to white or yellowish within 1–2 days of opening
- fruits capsules
- cultivated as outdoor ornamentals in frost-free areas and as tub plants elsewhere

Brunfelsia americana

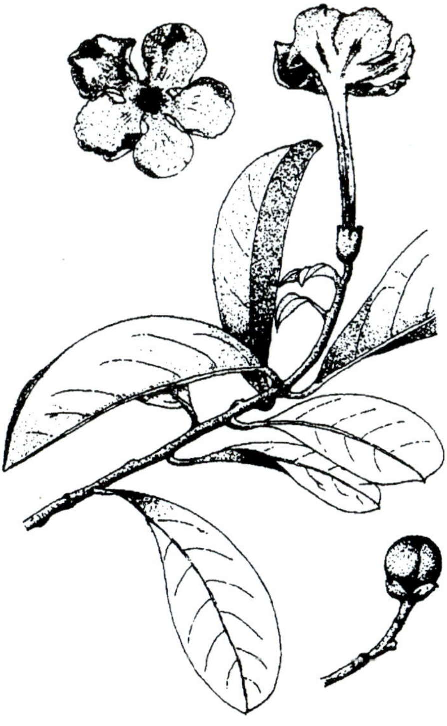

B. americana

X. Seizures: abrupt onset; hypersensitive to stimuli, ataxia, tetanic seizures

Calycanthus
sweetshrub
Calycanthaceae, strawberry-shrub family

Signs
- abrupt onset; muscle fasciculations, increased sensitivity to touch or other stimuli, ataxia
- seizures, tetanic or rigid, quickly unable to stand
- lateral recumbency, legs rigid, head twisted back
- third eyelid protrudes
- sometimes apparent sudden death

Pathology
- no distinctive lesions
- presence of plant fragments and cylindrical fruits in rumen

Treatment
- administer sedatives as for strychnine intoxication
- keep in quiet, dark area to reduce external stimuli and occurrence of seizures

Disease Problem
- mainly with ingestion of small amounts of fruits
- alkaloids; nonindole calycanthine and indoles chimonanthine, calycanthidine, and folicanthine
- strychninelike action; stimulate CNS, cause exaggerated reflexes

Taxonomy and Ecology
- two species of toxicologic concern:

C. floridus	Carolina allspice, sweetshrub, strawberry shrub
C. occidentalis	California allspice, California sweetshrub

- erect, aromatic, deciduous shrubs, 1–4 m tall
- leaves simple, opposite
- flowers aromatic with strawberry or pineapple scent, maroon, reddish-brown, or green tipped
- along streams and at edges of woods, eastern portion of continent and mountains of Pacific Coast states

Calycanthus floridus

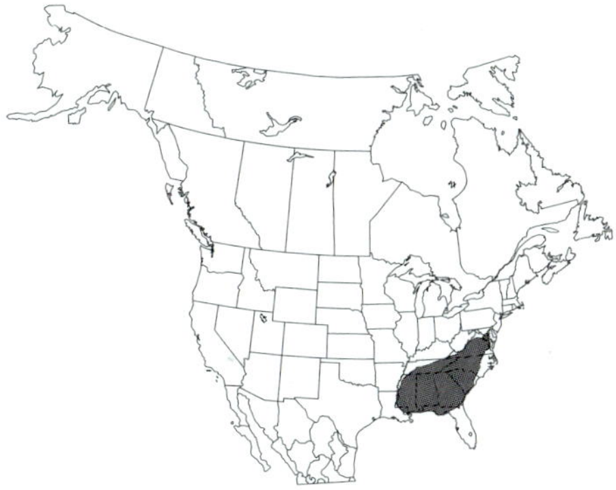

C. *floridus*

C. *occidentalis*

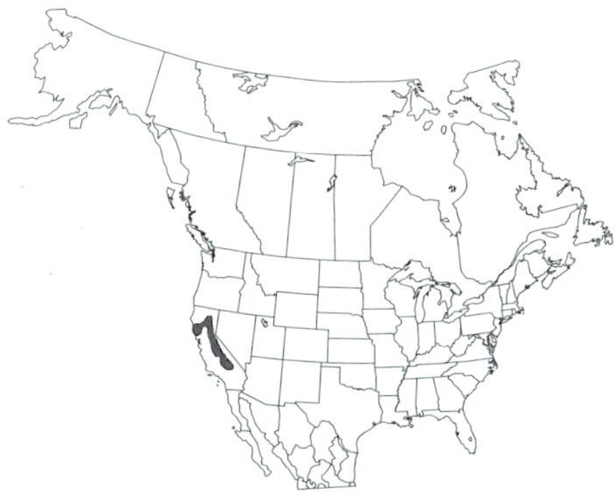

C. *occidentalis*

X. Seizures: abrupt onset; violent muscular activity, head and neck movements followed by seizures of entire body, falling with kicking/running leg movements

Cicuta
water hemlock
Apiaceae, carrot family

Signs

- violent muscular activity
- three stages: (1) uneasiness and twitching of muscles of lips, nose, face, and ears; (2) champing of jaws, grinding of teeth, throwing of head and neck grotesquely backward, sometimes frenzied activity and spasmodic diaphragmatic contractions may accompany the incipient seizures; (3) prolonged seizures, animal falls with kicking/running, tonic-clonic seizures, bellowing, frothy bloody salivation, bloat
- death may occur within 1–8 hours of ingestion, probably due to cardiopulmonary arrest
- increase in serum LDH, AST, and CK

Pathology

- no consistent lesions other than bruises, lacerations
- presence of root fragments in rumen, omasum, or stomach
- analysis of stomach contents for cicutoxin and cicutol

Treatment

- provide sedation; 30 mg/kg b.w. pentobarbital, 75 mg atropine
- administer activated charcoal orally

Disease Problem

- few bites of rootstalks, less than 0.5% b.w., quickly produce signs
- leaves of newly emerging plants may also present a risk
- effects typically last 3–4 hours; then death or recovery
- all animal species at risk, but especially herbivores
- acetylenic alcohols, cicutoxin, and cicutol, in oily yellowish liquid in chambers of rootstalk and lower stem
- toxin concentrations vary with season and geographical location; plants in areas with low toxicity potential also may have minimal yellowish liquid present in rootstalk
- reversible blockade of Na^+ and/or K^+ currents mainly on closed channels, similar to picrotoxin, a GABA inhibitor

Taxonomy and Ecology

- four species of toxicologic concern:

C. bulbifera	bulbous water hemlock
C. douglasii	Douglas water hemlock
C. maculata	water hemlock, beaver poison, snakeweed, spotted cowbane, children's bane, death-of-man, musquash root, poison parsley
C. virosa	Mackenzie's water-hemlock

- erect, perennial herbs, to 2.5 m tall
- stout hollow stems, lower stem and upper root chambered
- leaves 1–3-pinnately compound, veins extend to sinuses or notches between teeth
- flowers borne in umbels, small, petals 5, white
- typically along streams, near ponds or in marshy areas, across continent

Cicuta maculata

C. maculata

C. maculata

C. maculata

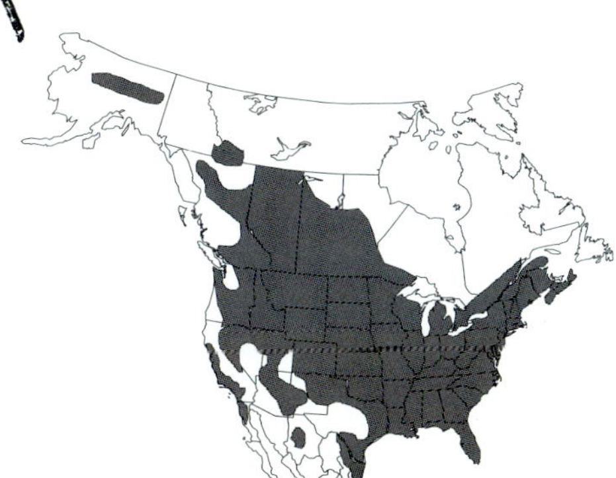

C. maculata

X. Seizures: abrupt onset; depression, increase in heart and respiratory rates, twitching, staggering, stiffness, collapse, paddling seizures, rigidity

Corydalis
corydalis
Fumariaceae, fumitory family

Signs

- appear abruptly and quickly become severe
- initially depression, then respiratory rate increases to >100/min, heart rate increases to a lesser degree
- followed quickly by apparent disorientation; muscular twitching especially of lips, face, eyelids; staggering; extreme stiffness especially of thoracic limbs
- animal may bite or chew at objects such as sticks or soil
- seizures may be precipitated by external stimuli, cause animal to fall accompanied by periodic episodes of paddling movements that terminate in rigidity
- as seizures worsen, respiration and heart rates become quite slow
- disease is usually of short duration and rapid recovery

Pathology

- few distinctive lesions

Treatment and Prevention

- provide sedation to control stiffness and seizures
- administer activated charcoal orally
- avoid grazing sheep in areas with large populations of plants

Disease Problem

- sporadic losses; sudden deaths of sheep and to a lesser extent cattle; goats least affected
- sparse, scattered populations of plants limit losses
- *C. caseana* is succulent, moisture content in excess of 95%, relished by livestock and grazed extensively
- hazardous throughout season, 3–4% b.w. may cause disease
- isoquinoline alkaloids such as phthalides, which act as GABA antagonists; possibly aporphines

Taxonomy and Ecology

- representative species of toxicologic concern:

C. aurea	golden corydalis, scrambled eggs, golden smoke, golden-fume-root
C. caseana	fitweed

- low-growing annual or perennial herbs, 5–150 cm tall
- leaves 1- or 2-pinnately compound, alternate
- flowers yellow, white, or pinkish, petals keeled
- fruits 2-valved capsules
- variety of soils and habitats; typically small populations, sometimes large in disturbed sites

Corydalis aurea

C. aurea

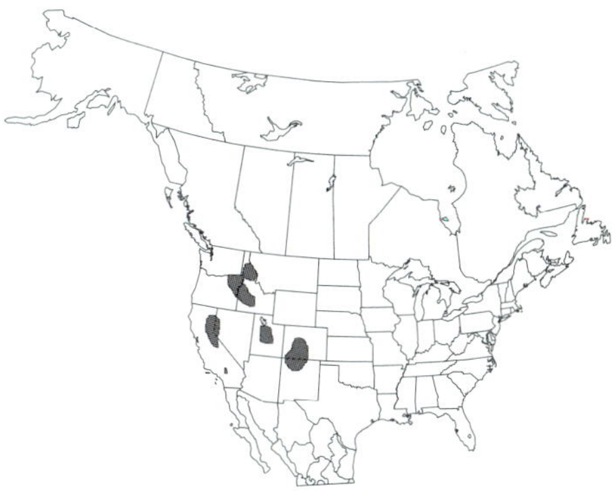

C. aurea

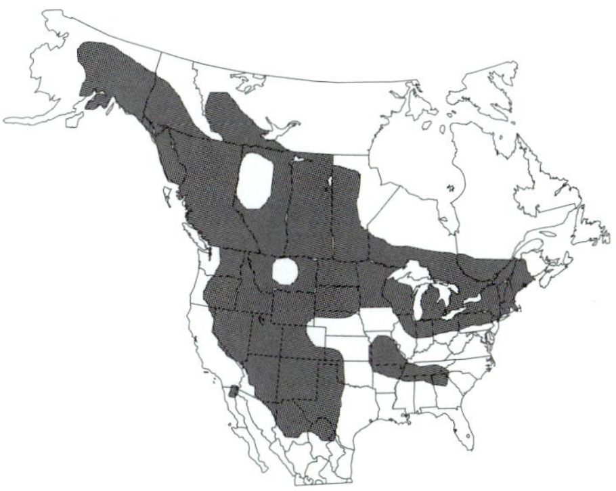

C. aurea

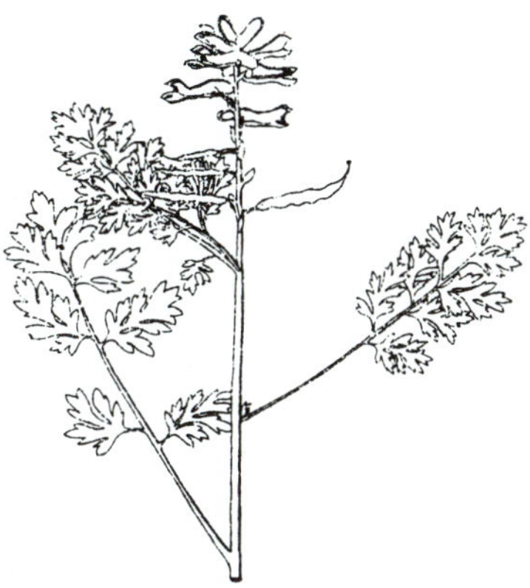

C. caseana

C. caseana

X. Seizures: abrupt onset; weakness, convulsive movements of head and limbs, swallowing difficulty

Gelsemium
jessamine
Loganiaceae, logania family

Signs
- abrupt onset; muscular weakness, clonic convulsive movements of head and thoracic and pelvic limbs
- decrease in respiration rate and temperature
- swallowing difficulty
- vision problems
- respiratory paralysis

Pathology
- no distinctive lesions

Treatment
- administer activated charcoal orally
- provide respiratory assistance

Disease Problem
- all livestock species at risk, including poultry
- effects cumulative and may show signs several days after removal from plant access
- evergreen and high risk in winter
- all plant parts toxic
- alkaloids, gelsemine, and sempervirine
- paralysis of nerve endings

Taxonomy and Ecology
- two species of toxicologic concern:

G. rankinii	evening trumpet-flower
G. sempervirens	yellow jessamine, Carolina jessamine, woodbine

- perennial, evergreen, twinning, woody vines
- flowers showy, funnelform, petals 5, fused, yellow
- fruits flattened capsules
- cultivated as climbing vine in southern half of continent; open woods, thickets, swamps, low areas, and open fields in southeastern portion of continent
- *G. rankinii* rare on southeastern coastal plains

Gelsemium sempervirens

G. sempervirens

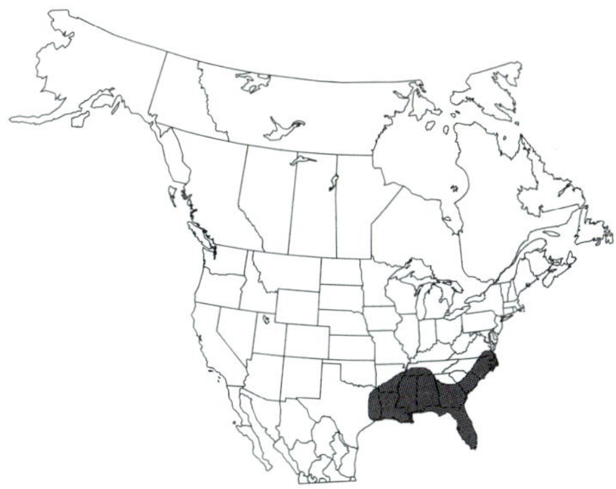

G. sempervirens

G. sempervirens

X. Seizures: abrupt onset; sheep mainly; ataxia, nervousness, weakness, muscle twitching, colic, excitability, excessive salivation, tetanic seizures

Lupinus
lupine
Fabaceae, pea family

Signs
- depression, ataxia, muscle weakness and twitching, nervousness, breathing difficulty
- sheep: become quite excitable, with head pressing, head butting, frenzy, excessive salivation, tetanic seizures
- horses: colic, exaggerated walk with legs lifted high

Pathology
- no distinctive lesions

Treatment and Prevention
- provide relief of neurologic signs
- move sheep slowly through stand of lupines in fruit so they can graze selectively
- avoid overnight bedding near lupines

Disease Problem
- "acute lupine toxicity," mainly sheep
- ingestion of mature plants bearing legumes, fresh or dry
- 1% b.w. may cause intoxication
- horses and cattle low risk
- quinolizidine, and possibly indole and piperidine-derived alkaloids; concentrations especially high in seeds
- agonist effects on central and peripheral nicotinic acetylcholine receptors
- also affects reproductive system (teratogenic effects)

Taxonomy and Ecology
- representative species of toxicologic concern:

L. arbustus	spur lupine
L. argenteus	silvery lupine, tail-cup lupine, grassland lupine
L. latifolius	broad-leaved lupine
L. leucophyllus	woolly-leaved lupine, western lupine, velvet lupine, white-leaved lupine
L. onustus	woodland lupine, Plumas lupine
L. sericeus	silky lupine, sink lupine
L. sulphureus	sulfur lupine

- erect herbs or subshrubs, to 1 m tall
- leaves palmately compound, alternate, leaflets—3–17, typically 5
- inflorescences terminal racemes or spikes
- flowers pealike, petals 5, blue, violet, lilac, crimson, white, or yellow
- legumes flattened, seeds 2–12
- wide range of habitats; grasslands, shrublands, open forests; greatest abundance in western half of continent

Lupinus latifolius

L. leucophyllus

L. argenteus

L. leucophyllus

X. Seizures: usually abrupt onset; retching, vomiting, excessive salivation, diarrhea, weakness, ataxia, seizures with tetanic or paddling leg movements

Melia
chinaberry
Meliaceae, mahogany family

Signs

- increased salivation, retching, vomiting, anorexia, diarrhea; occur several hours after ingestion
- early excitement, later paralysis, digestive disturbance
- respiration labored and irregular
- weakness, muscle spasms, ataxia, tetanic- or clonic-type seizures
- disease usually resolved in 24–48 hours, recovery or death

Pathology

- focal edema and reddening of mucosa of stomach and small intestine
- gaseous distension of stomach
- presence of drupes or pits in stomach

Treatment

- induce emesis prior to onset of signs
- perform gastric lavage and/or give activated charcoal orally
- provide sedation
- administer fluids and immediate-acting corticosteroids

Disease Problem

- dogs, pigs, ruminants, rabbits, and most other animal species
- fruits, bark, and leaves toxic, possibly flowers
- toxic dosage is 0.2–0.3% b.w., more for ruminants
- fruits yellow and sweet when ripe
- triterpene epoxides, meliatoxins isolated from fruits
- also affects digestive system

Taxonomy and Ecology

- one introduced species of toxicologic concern:

M. azedarach chinaberry, Persian lilac, white cedar, pride-of-India, Indian lilac

- deciduous tree, to 15 m tall
- leaves 2-pinnately compound, leaflets 50 or more
- flowers borne in panicles, petals 5–6, white, pinkish lavender, or purple
- fruits globose, yellow drupes
- cultivated as ornamental in southern portion of continent; escaped cultivation and naturalized

Melia azedarach

M. azedarach

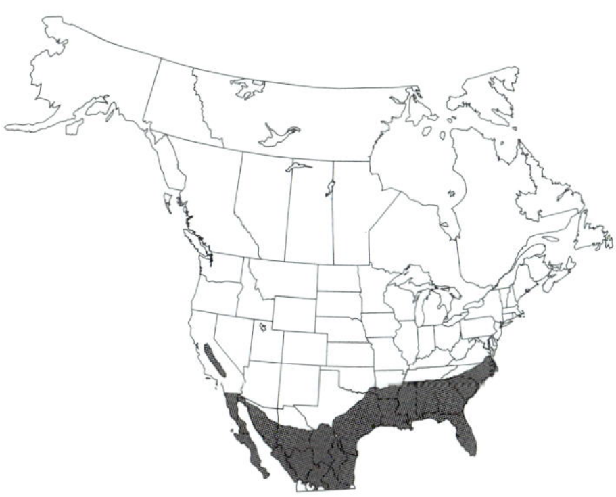

M. azedarach

X. Seizures: usually abrupt onset; cattle mainly; seizures with stiffness, trembling, collapse

Sophora
mescal bean
Fabaceae, pea family

Signs
- cattle: stiffness of pelvic limbs, trembling of rump and shoulder muscles, falling and unable to rise
- sheep and goats: brief somnolence, then likely able to walk
- forced exercise may precipitate occurrence of signs

Pathology
- no specific lesions

Treatment
- provide relief of symptoms

Disease Problem
- mainly cattle, sheep, goats
- pets may also be at risk if seeds are well chewed
- less than 1% b.w. of seeds and foliage (especially mature) produce both neurologic and cardiac effects, especially after forced exercise
- acute: signs appear after ingestion of plant for short time; calves may die in a few hours or days
- small amounts eaten for a week or more not cumulative
- quinolizidine alkaloids of two types: neurotoxic nicotinelike tricyclics and neurotoxic/cardiotoxic tetracyclics
- also affects cardiovascular system

Taxonomy and Ecology
- only 1 of 12 native and introduced species definitely toxic:

S. secundiflora	mescal bean, coralbean

- evergreen shrub or tree, to 3–4 m tall
- leaves 1-pinnately compound, alternate, leaflets 5–13
- flowers pealike, petals 5, blue-violet
- fruits woody legumes, markedly constricted between showy, red seeds
- across southern Texas into New Mexico and Mexico

Sophora secundiflora

S. secundiflora

S. secundiflora

S. secundiflora

S. secundiflora

X. Seizures: abrupt onset; ataxia, weakness, depression, possible hypothermia, spasmodic muscular activity, seizures with paddling leg movements

Xanthium
cocklebur
Asteraceae, sunflower family

Signs and Diagnostic Aids
- pigs: abrupt onset, depression, weakness, loss of appetite, vomiting, tucked-up appearance, ataxia, spasmodic muscular activity, recumbency
- convulsive paddling, coma, and death within a few hours or several days, possibly hypothermia, peculiar gait or posture with head held erect
- calves: similar but no vomiting and delayed onset
- if prominent signs appear, almost invariably fatal
- tenfold increase in serum AST, SDH
- sharp decline in blood glucose levels to as low as 10 mg/dl

Pathology
- protein-rich fluid in abdominal cavity, pericardial sac, and thoracic cavity
- gelatinous edema of gall bladder wall, hepatic capsule, and subserosa of small intestine
- liver: swollen, accentuated markings on cut surfaces
- microscopically: severe, diffuse centrilobular hepatocellular necrosis involving almost entire lobules
- presence of plant fragments or intact sprouts in stomach
- assay rumen contents and urine for carboxyatractyloside

Treatment
- parenteral glucose and bicarbonate, but not curative
- milk, lard, or other fatty substances somewhat protective in pigs; reduce absorption of toxin
- administer activated charcoal orally early
- administer neostigmine or physostigmine IM for temporary alleviation

Disease Problem
- seeds or new sprouts or cotyledons
- all animal species susceptible, about 1% b.w. of sprouts
- spring and early summer; moisture and warm temperatures cause massive sprouting
- appearance mainly in sandy soils adjacent to water in low areas, plants sprout when water recedes
- novel diterpene glycoside, carboxyatractyloside (CAT) concentrated in seeds within bur and 2-leaf seedling stage
- inhibition of oxidative phosphorylation by blocking transport of phosphoryl groups across mitochondrial membrane, effects antagonized by ADP
- also affects liver

Taxonomy and Ecology
- two species of toxicologic concern:

X. spinosum	spiny cocklebur, clotbur
X. strumarium	common cocklebur, burweed, abrojo, sheep bur, bur thistle, ditchbur

- annual herbs from taproots, some armed with spines, 20–200 cm tall
- leaves simple, alternate
- flowers (florets) inconspicuous, borne in staminate and pistillate heads
- only disk florets present
- achenes enclosed in brown prickly burs
- disturbed sites, often forming large, dense populations in floodplains, along shorelines of ponds and reservoirs

Xanthium strumarium

X. strumarium

X. strumarium

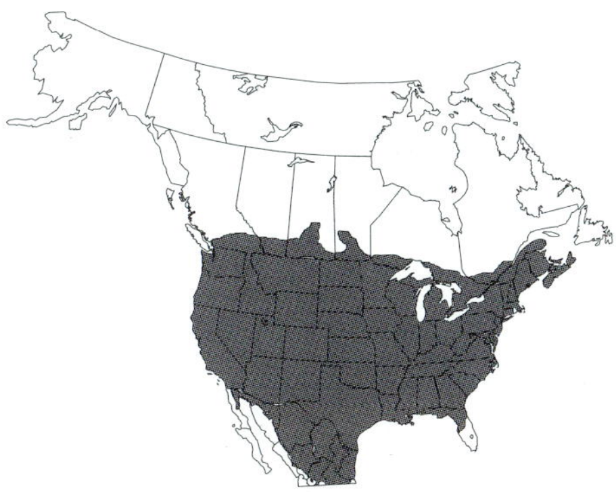

X. strumarium

XI. Seizures: abrupt onset; sudden death; ruminants mainly; distress, ataxia, weakness, collapse with paddling and tetanic seizures

Prunus
cherries
Rosaceae, rose family

Signs
- sudden onset of apprehension and distress
- quickly followed by weakness, ataxia, labored respiration, collapse, and lateral recumbency with periodic paddling and tetanic seizures
- sequence of signs in 5–15 minutes, terminating in death or recovery

Pathology
- no distinctive changes

Treatment
- seldom available soon enough to be effective
- administer sodium thiosulfate i.v. 0.25–0.5 g/kg b.w.
- sodium nitrite, 10 mg/kg b.w. i.v helpful in combination with thiosulfate, but not required

Disease Problem
- many species cyanogenic, but most not high risk
- spring foliage of *P. laurocerasus*, *P. serotina*, and *P. virginiana* pose greatest risks
- rare in monogastric species, but some risk for dogs
- seeds toxic, not pulp of fruit
- cyanogenic glycoside, prunasin, hydrolyzed to α-hydroxynitrile aglycone, then to ρ-hydroxybenzaldehyde and hydrocyanic or prussic acid (HCN)
- some hydrolysis occurs with insect or frost damage and wilting but most takes place in microbe rich rumen
- cyanide affects enzyme systems throughout the body, especially noteworthy inhibition of cytochrome oxidase pathway in brain
- also affects blood

Taxonomy and Ecology
- representative species of particular toxicologic concern:

P. caroliniana	laurel cherry, mock orange
P. laurocerasus	cherry-laurel
P. pensylvanica	bird cherry, pin cherry, wild red cherry
P. serotina	black cherry, wild cherry
P. virginiana	common chokecherry

- shrubs or trees, spur branches, conspicuous lenticels
- leaves simple, alternate, deciduous, glands 1 or 2 on petioles on margins of blade bases
- flowers showy, produced before, with, or after leaves
- petals 5 or 10, spreading, white, pink, red, or greenish
- fruits fleshy drupes
- variety of habitats across continent

Prunus serotina

P. serotina

P. virginiana

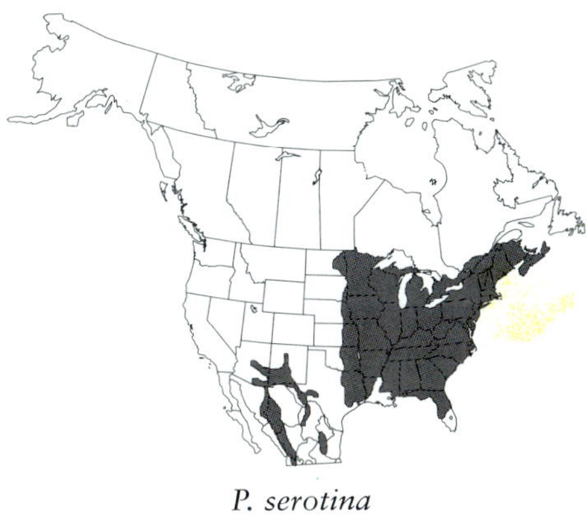

P. serotina

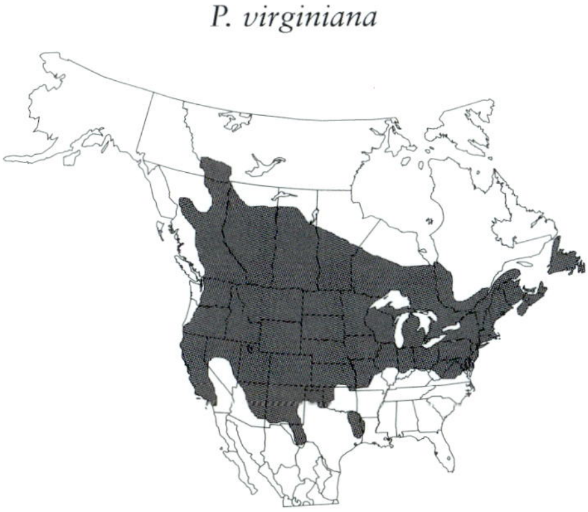

P. virginiana

XI. Seizures: abrupt onset; sudden death; ruminants mainly; distress, ataxia, weakness, collapse with paddling and tetanic seizures

Sorghum
sorghum, sudan
Poaceae, grass family

Signs
- sudden onset of apprehension and distress
- quickly followed by weakness, ataxia, labored respiration, collapse, and lateral recumbency with periodic paddling and tetanic seizures
- sequence of signs in 5–15 minutes, terminating in death or recovery

Pathology
- no distinctive changes

Treatment and Prevention
- seldom available soon enough to be effective
- administer sodium thiosulfate i.v. 0.25–0.5 g/kg b.w.
- sodium nitrite, 10 mg/kg b.w. i.v helpful in combination with thiosulfate but not required
- forage should not be grazed until it is >45–60 cm high

Disease Problem
- cattle: signs may occur minutes or hours after animals begin grazing new *Sorghum* pasture; lethal quantities of *S. halepense* sometimes consumed in 15 minutes
- rate of growth, stage of growth, moisture, temperature influence toxicity
- grain sorghums highest in HCN potential, Sudan grass lowest, and hybrids intermediate
- when plants 2.5–15 cm tall, risk is quite high, but declines markedly as height exceeds 20 cm; danger very low when Sudan grass has gone to seed
- cyanogenic glucoside, dhurrin, hydrolyzed to an α-hydroxynitrile aglycone, then to hydrocyanic or prussic acid (HCN)
- some hydrolysis occurs with insect or frost damage and wilting but most takes place in microbe-rich rumen
- cyanide affects enzyme systems throughout the body, especially noteworthy inhibition of cytochrome oxidase pathway in brain
- also causes: photosensitization, nitrate intoxication, and, in horses, urinary incontinence, cystitis, ataxia, and possibly teratogenic effects when grazed for long periods

Taxonomy and Ecology
- three introduced species of toxicologic concern:

S. x *almum*	sorghum almum, Columbus grass
S. bicolor	sorghum, shattercane, sudan, milo, millet, Sudan grass
S. halepense	Johnsongrass

- erect annuals or perennials, 0.6–3 m tall
- leaves flat, 12–120 mm wide
- inflorescences panicles of paired spikelets
- spikelets dorsally compressed; glumes 2, equal, leathery
- cultivated for forage, hay, silage, and grain
- *S. halepense* is weed of roadsides, ditches, and other disturbed sites

Sorghum halepense

S. halepense

S. halepense

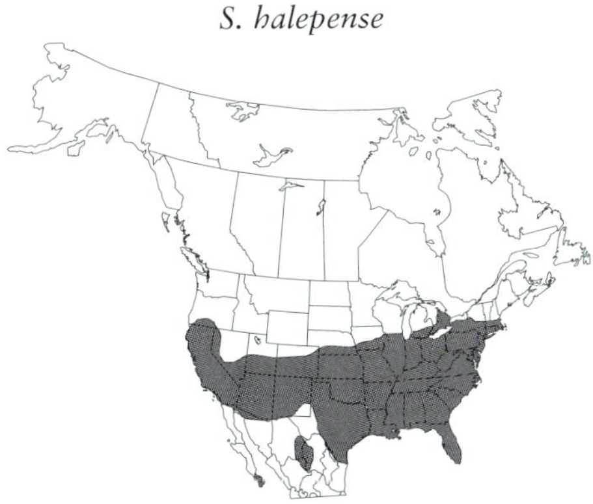

S. halepense

XI. Seizures: abrupt onset; sudden death; ruminants mainly; distress, ataxia, weakness, collapse with paddling and tetanic seizures

Suckleya
poison suckleya
Chenopodiaceae, goosefoot family

Signs

- sudden onset of apprehension and distress
- quickly followed by weakness, ataxia, labored respiration, collapse, and lateral recumbency with periodic paddling and tetanic seizures
- sequence of signs in 5–15 minutes, terminating in death or recovery

Pathology

- no distinctive changes

Treatment

- seldom available soon enough to be effective
- administer sodium thiosulfate i.v. 0.25–0.5 g/kg b.w.
- sodium nitrite, 10 mg/kg b.w. i.v helpful in combination with thiosulfate but not required

Disease Problem

- sporadic cause of death in sheep and cattle
- typically a problem when dense stands of plants are established around edges of shallow stock ponds as water slowly recedes
- cyanogenic glucoside, dhurrin, hydrolyzed to an α-hydroxynitrile aglycone, then to hydrocyanic or prussic acid (HCN)
- some hydrolysis occurs with insect or frost damage and wilting but most takes place in microbe-rich rumen
- cyanide affects enzyme systems throughout the body, especially noteworthy inhibition of cytochrome oxidase pathway in brain
- risk of intoxications increased if animals drink shortly before or after ingestion of plant

Taxonomy and Ecology

- one species of toxicologic concern:

  S. suckleyana poison suckleya

- annual fleshy herb, often reddish, prostrate or ascending stems, 10–50 cm long
- leaves simple, alternate
- flowers staminate and pistillate; in axillary glomerules
- dense populations in low areas, around shallow ponds, dry lake beds, edges of lakes, and around water tanks; primarily eastern front of Rocky Mountains

Suckleya suckleyana

S. suckleyana

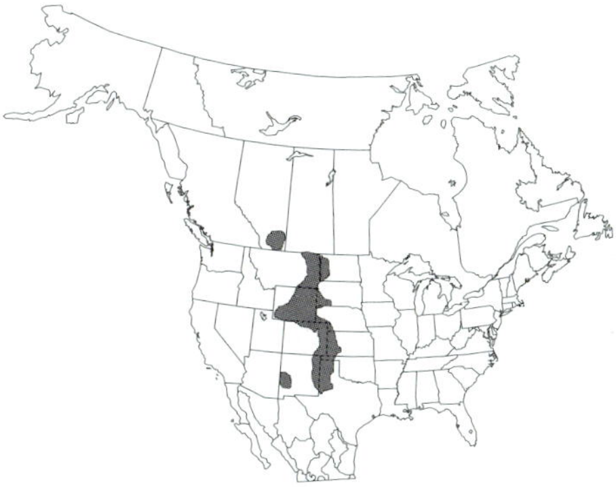

S. suckleyana

S. suckleyana

XI. Seizures: abrupt onset; sudden death; ruminants mainly; distress, ataxia, weakness, collapse with paddling and tetanic seizures

Triglochin
arrowgrass
Juncaginaceae, arrowgrass family

Signs
- sudden onset of apprehension and distress
- quickly followed by weakness, ataxia, labored respiration, collapse, and lateral recumbency with periodic paddling and tetanic seizures
- sequence of signs in 5–15 minutes, terminating in death or recovery

Pathology
- no distinctive changes

Treatment
- seldom available soon enough to be effective
- administer sodium thiosulfate i.v. 0.25–0.5 g/kg b.w.
- sodium nitrite, 10 mg/kg b.w. i.v helpful in combination with thiosulfate but not required

Disease Problem
- succulent, remains semigreen and toxic in hay; readily eaten by sheep and cattle fresh or dried
- 1% b.w. fresh or 0.25% b.w. dried may result in intoxication
- tyrosine-derived cyanogenic glucoside, triglochinin hydrolyzed to α-hydroxynitrile aglycone, then to hydrocyanic or prussic acid (HCN)
- some hydrolysis occurs with insect or frost damage and wilting but most takes place in microbe-rich rumen
- cyanide affects enzyme systems throughout the body, especially noteworthy inhibition of cytochrome oxidase pathway in brain

Taxonomy and Ecology
- four species of toxicologic concern:

T. gaspensis	gaspe arrowgrass
T. maritima	seaside arrowgrass, low arrowgrass, graceful arrowgrass
T. palustris	marsh arrowgrass, small arrowgrass
T. striata	three-ribbed arrowgrass, southern arrowgrass

- perennial herbs from rhizomes, 20–80 cm tall
- leaves simple, alternate, basal, fleshy, crescent-shaped in cross section
- flowers in spikes or racemes
- grow in tufts, sometimes forming rather dense and/or extensive patches in marshy sites, along the edges of brackish waters, and in depressions in areas of wet saline marshland soils; across northern half of continent

Triglochin maritima

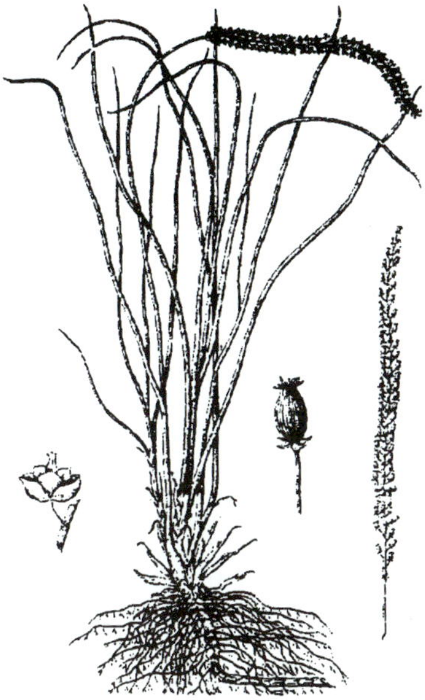

T. palustris

T. maritima

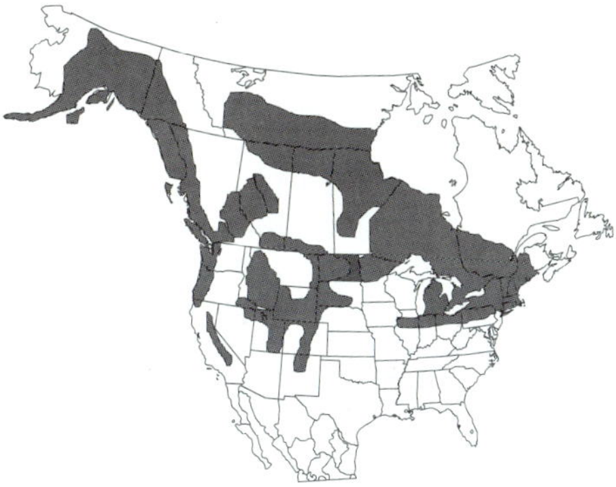

T. palustris

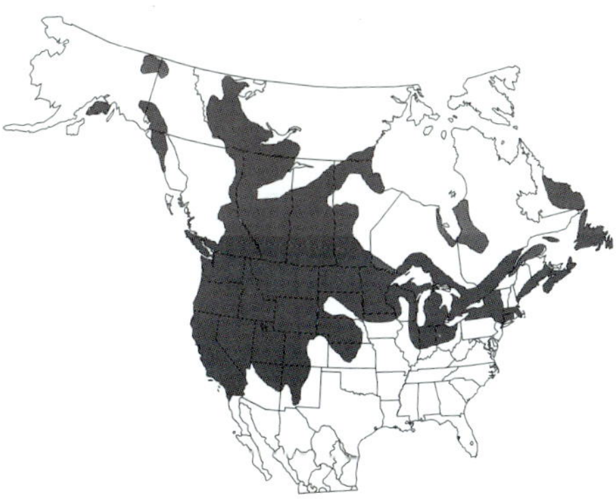

T. maritima

XII. Tremors: beginning with neck and shoulder muscles; cattle mainly; ears erect, nervous, stiff, wobbly gait, swaying on pelvic limbs; collapse

Cynodon
Bermudagrass
Poaceae, grass family

Signs
- increased nervousness, seizures with ears erect, fine tremor of head and neck to shoulders
- move with stiff, wobbly, swaying movement of rear legs; collapse, head twisted backward
- usually able to arise after a brief rest only to tremor and fall again
- low case fatality rate; deaths mainly due to misadventure

Pathology
- no distinctive changes

Treatment
- if possible, remove from pasture
- if recumbent, provide water and feed
- avoid excessive stimulation of animal in hazardous areas where falling may cause injury or drowning

Disease Problem
- "Bermudagrass staggers"; associated with mature stands and hay
- rare; mainly cattle; mild effects in horses and goats
- possibly ergot-type alkaloids from fungal endophytes
- toxin not transmitted in milk
- also affects respiratory system and skin

Taxonomy and Ecology
- one species of toxicologic concern:

  C. dactylon Bermudagrass, common or coastal

- stoloniferous and usually rhizomatous perennial, 4–50 cm tall
- leaves cauline
- inflorescences digitate whorl of 1-sided spikes of spikelets
- spikelets laterally compressed, florets 1
- cultivated forage across southern half of continent

Cynodon dactylon

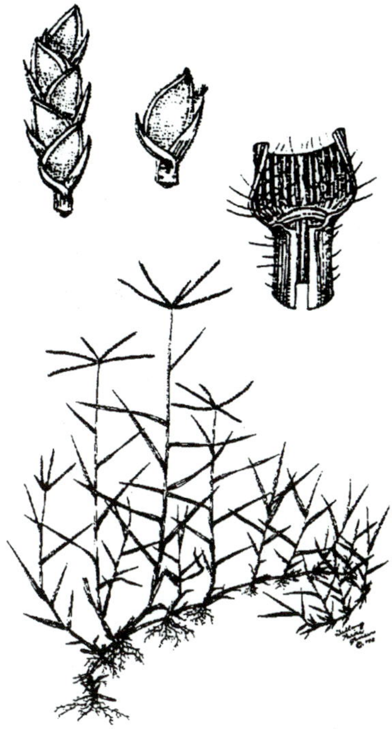

C. dactylon

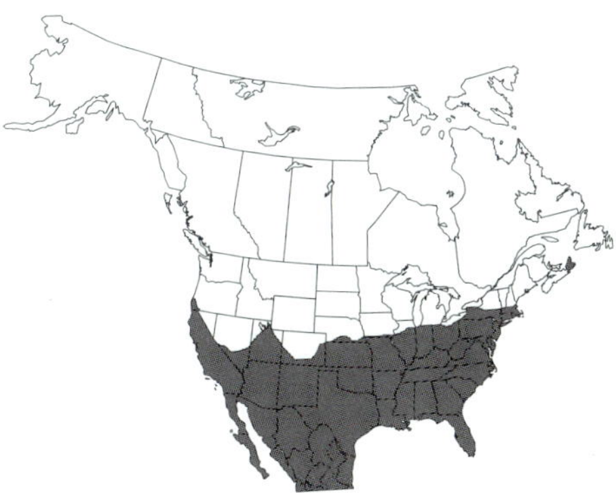

C. dactylon

XII. Tremors: beginning with neck and shoulder muscles; cattle mainly; ears erect, nervous, stiff, wobbly gait, swaying on pelvic limbs; collapse

Lolium
perennial ryegrass
Poaceae, grass family

Signs
- increased nervousness, seizures with ears erect, fine tremor of head and neck to shoulders
- move with stiff, wobbly, swaying movement of rear legs; collapse, head twisted backward
- usually able to arise after a brief rest only to tremor and fall again
- low case fatality rate; deaths mainly due to misadventure

Pathology
- no distinctive changes

Treatment
- if possible, remove from pasture
- if recumbent, provide water and feed
- avoid excessive stimulation of animal in hazardous areas where falling may cause injury or drowning

Disease Problem
- "ryegrass staggers"; associated with mature stands
- uncommon; mainly in summer and fall
- lolitrem alkaloids from *Neotyphodium lolii*, fungal endophyte
- also causes liver disease with photosensitization and summer slump similar to that caused by *Festuca*

Taxonomy and Ecology
- one species of toxicologic concern:

L. perenne	perennial ryegrass, English ryegrass, Italian ryegrass

- perennial, 30–100 cm tall
- ligules membranous
- inflorescences 2-sided spikes of spikelets
- spikelets laterally compressed, florets 2–25, glumes 1
- cultivated forage crop; across continent, especially in north

Lolium perenne

L. perenne

L. perenne

XII. Tremors: beginning with neck and shoulder muscles; cattle mainly; ears erect, nervous, stiff, wobbly gait, swaying on pelvic limbs; collapse

Paspalum
paspalum
Poaceae, grass family

Signs
- increased nervousness, seizures with ears erect, fine tremor of head and neck to shoulders
- move with stiff, wobbly, swaying movement of rear legs; collapse, head twisted backward
- usually able to arise after a brief rest only to tremor and fall again
- low case fatality rate; deaths mainly due to misadventure

Pathology
- no distinctive changes

Treatment
- if possible, remove from pasture
- if recumbent, provide water and feed
- avoid excessive stimulation of animal in hazardous areas where falling may cause injury or drowning

Disease Problem
- "paspalum staggers"; associated with mature plants with caryopses/seeds present
- hazard increases with plant maturity especially after rainy/humid weather
- cattle mainly; sheep also susceptible
- fungal endophyte, *Claviceps paspali*
- invades developing plant ovaries, forms sclerotia containing neurotoxic indolic tremorgens, paspalinine, paspalitrems
- inhibition of GABA- and glycine-inhibitory pathways
- paspalinines appear to be eliminated only by biliary excretion with possible enterohepatic recycling perpetuating activity

Taxonomy and Ecology
- two introduced species of toxicologic concern:

P. dilatatum	dallisgrass, large watergrass
P. notatum	Bahia grass

- annuals or perennials, 10–200 cm tall
- inflorescences 1-sided racemes of spikelets
- spikelets dorsally compressed, florets 2, glume 1
- cultivated forages; escaped and naturalized

Paspalum dilatatum

P. dilatatum

P. dilatatum

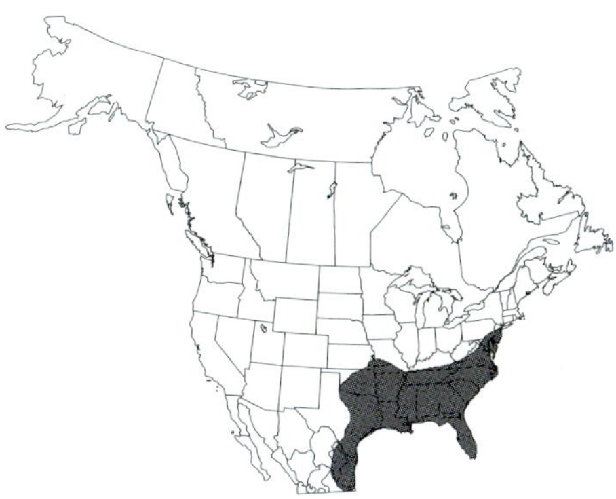

P. dilatatum

XII. Tremors: beginning with neck and shoulder muscles; sheep mainly; ears erect, nervous, excitable, ataxic, stiff, shaking; collapse

Phalaris
canarygrass
Poaceae, grass family

Signs and Diagnostic Aids

- abrupt or slow onset; tremors, excitement, ataxia, stiff gait, shaking, collapse; or recurrent episodes of head nod, ataxia, weakness
- high case fatality rate; deaths continue for several days after removal from pasture
- examine CSF: increase in protein and small mononuclear cells, presence of greenish intracytoplasmic granules in macrophages

Pathology

- focal, bilateral greenish-gray-blue discoloration of various sites in brain
- microscopically: granular pigment in neurons of gray matter, spinal neuronal degeneration

Treatment and Prevention

- none available
- possibly prevent with drench of 7 mg cobalt/animal as chloride or sulfate

Disease Problem

- "staggers syndrome"; associated only with green plants
- mainly sheep; less problem in cattle
- onset usually after several weeks or months of grazing
- periodic seizures, may begin after animals are moved off pasture
- possibly due to tryptamine alkaloids, DMT, 5-MDMT, or β-carboline alkaloids
- agonists of serotonin receptors or inhibit monoamine oxidase
- also may affect the cardiovascular system (sudden death) after grazing for only a few days or rarely cause polioencephalomalacia (PEM) (page 160)

Taxonomy and Ecology

- six species of possible toxicologic concern:

P. angusta	timothy canarygrass
P. aquatica	harding grass, toowoomba
P. arundinacea	reed canarygrass, ribbon-grass
P. canariensis	canarygrass, birdseed grass
P. caroliniana	Carolina canarygrass, southern canarygrass
P. minor	littleseed canarygrass

- annuals or perennials, 10–200 cm tall
- leaves cauline, blades flat, ligules membranous
- inflorescences spicate panicles of spikelets
- spikelets laterally compressed, fertile florets 1, glumes 2, longer than florets
- cultivated forages; introduced and native species

Phalaris caroliniana

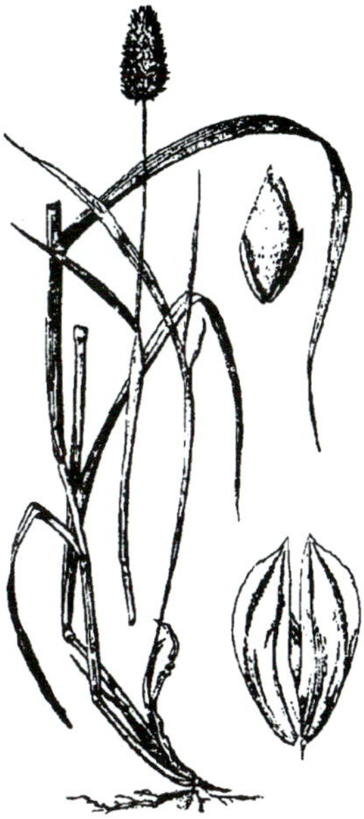

P. canariensis

P. caroliniana

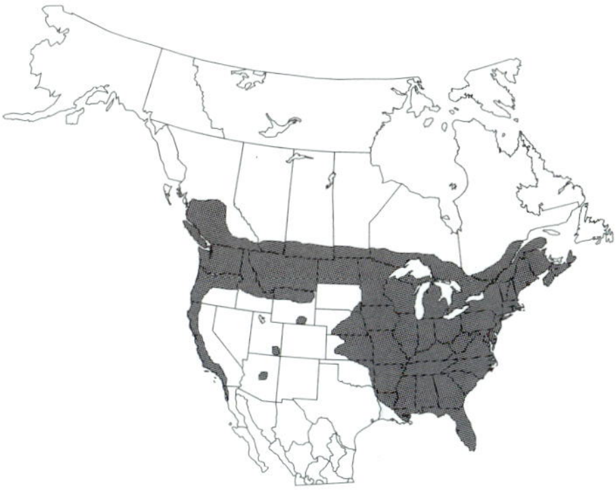

P. canariensis

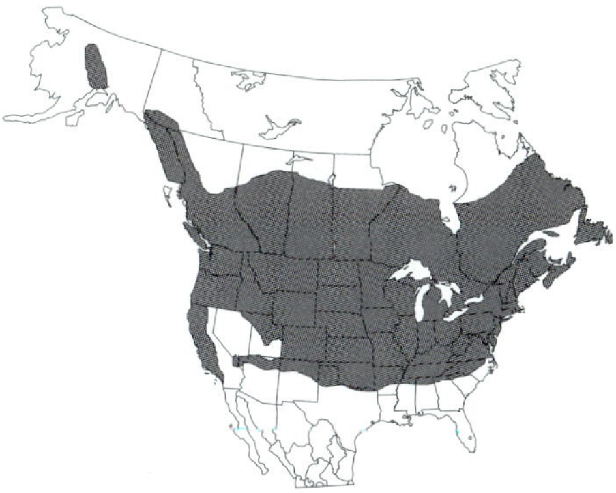

P. arundinacea

XIII. Tetany and Generalized Tremors: early signs of weakness, nervousness, ataxia, excessive salivation; later signs of severe depression, recumbency, hypothermia, slow heart and respiration rates

Conium
poison hemlock
Apiaceae, carrot family

Signs
- early phase—slight nervousness, tremors, muscular weakness, ataxia, excess salivation and lacrimation, groaning, increased frequency of urination, colic
- later phase—severe depression and/or narcosis, progressive paresis leading to recumbency, bradycardia, hypothermia, respiratory depression
- typically, signs subside, animal recovers in 8 hours

Pathology
- no distinctive lesions
- "mousey" odor when rumen opened

Treatment
- administer activated charcoal orally
- atropine affects only excess salivation, lacrimation

Disease Problem
- CNS disease; mainly cattle, pigs
- contamination of green-chop alfalfa or hay with stems and leaves
- because dense stands may occur along roadsides, hay cut in these areas may present a special risk for several weeks after cutting
- toxic dosage ranges from 0.2–0.8% b.w. depending on animal species; cattle most susceptible
- volatile pyridine alkaloids, coniine, N-methylconiine, γ-coniceine
- stimulate and block autonomic ganglia; larger doses; neuromuscular blockade similar to effect of nicotine
- also affects urinary and reproductive systems; fetal malformations in cattle and pigs

Taxonomy and Ecology
- one species of toxicologic concern

C. maculatum	poison hemlock, spotted hemlock, stinkweed, poison parsley

- erect annual or biennial, stout hollow stems, large white taproot, to 3 m tall
- leaves 2–4-pinnately compound, leaflets finely dissected
- flowers borne in compound umbels, petals white
- moist, often disturbed sites; sometimes large dense stands, across continent except for deserts; abundant in Great Plains

Conium maculatum

C. maculatum

C. maculatum

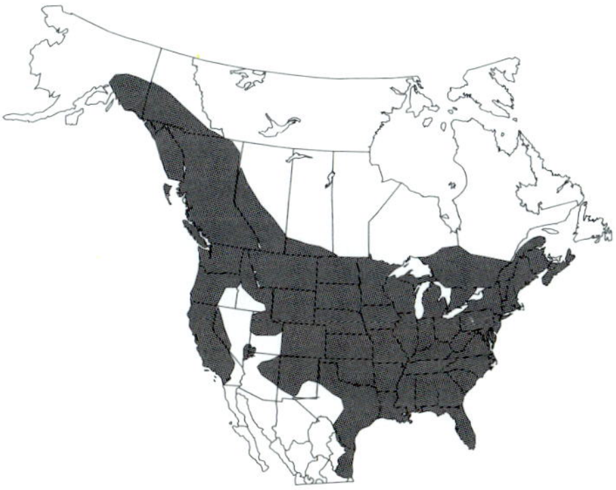

C. maculatum

XIII. Tetany and Generalized Tremors: weight loss, reluctance to move, weakness, muscular tremors when forced to move, joint stiffness, acetone odor to breath

Eupatorium
white snakeroot
Asteraceae, sunflower family

Signs
- cattle: marked weight loss, listlessness, reluctance to move
- later, weakness (lying down most of time), muscular tremors when forced to stand or move
- constipation, apparent joint stiffness and occasional belligerence
- acetone odor, especially breath
- signs may appear in nursing calves, without signs in cow
- sheep and pigs: shorter disease, severe
- horses: swallowing difficulty, slobbering, dyspnea, weak unsteady gait, constipation, patchy sweating, trembling, severe weakness, choke
- horses: cardiac effects, irregular rhythm, jugular vein distension and pulsation, ventral edema, AV block, premature ventricular contractions, amplitude changes, increase in LDH and CPK
- goats and sometimes sheep: loss of appetite, depression, head pressing, paddling seizures and photosensitization

Pathology
- liver: pale, gall bladder distended with bile
- horses: pale grayish myocardial streaks with ventricular dilation
- goats: severe centrilobular hepatic necrosis

Treatment
- provide nursing care, symptomatic treatment, good forage
- administer parenteral carbohydrates for low blood glucose levels
- administer activated charcoal orally, prevent enterohepatic cycling

Disease Problem
- "trembles" in animals; "milk sickness" in humans
- cumulative effect; large amounts for few days or small amounts for several weeks or more
- 5–10% b.w. may produce intoxication
- when dried in hay, may retain intoxication potential for days or weeks
- trematol; mixture of sterols and derivatives of methyl ketone benzofuran, tremetone, and derivatives
- impairment of TCA cycle, decreased utilization of glucose; enzyme inhibition resulting in acidosis and ketonemia
- tremetol fat-soluble, excreted in milk
- also affects cardiovascular system and liver

Taxonomy and Ecology
- one species of toxicologic concern:

E. rugosum	white snakeroot, rich weed, white sanicle

- now classified as *Ageratina altissima*
- erect, perennial herb, 30–150 cm tall
- leaves simple, opposite, conspicuously 3-ribbed
- flowers (florets) borne in numerous, small heads
- disk florets only, white
- moist woodlands of eastern half of continent

Eupatorium rugosum

E. rugosum

E. rugosum

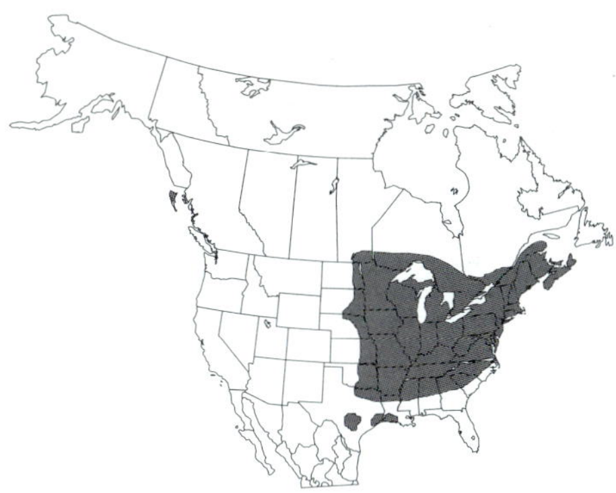

E. rugosum

XIII. Tetany and Generalized Tremors: weight loss, reluctance to move, weakness, muscular tremors when forced to move, joint stiffness, acetone odor to breath

Isocoma
goldenbush
Asteraceae, sunflower family

Signs
- depression, weakness (lying down most of the time), muscular tremors when forced to stand or move
- constipation, apparent joint stiffness and occasional belligerence
- acetone odor, especially breath
- sheep: humped stance

Pathology
- liver pale, gall bladder distended

Treatment
- provide nursing care, symptomatic treatment, good forage
- administer parenteral carbohydrates for low blood glucose levels
- administer activated charcoal orally, prevent enterohepatic cycling

Disease Problem
- "alkali disease" or "milk sickness"
- cumulative effect; large amounts of plant, 1–2% b.w. daily for a few days or small amounts for several weeks
- ruminants mainly; metabolic derangement
- trematol; mixture of sterols and derivatives of methyl ketone benzofuran, tremetone, and its derivatives
- impairment of TCA cycle, decreased utilization of glucose; enzyme inhibition resulting in acidosis and ketonemia
- tremetol fat-soluble, excreted in milk

Taxonomy and Ecology
- three species of toxicologic concern:

I. coronopifolia	common goldenbush
I. pluriflora	jimmy weed, rayless goldenrod
I. tenuisecta	burroweed

- perennial subshrubs, erect, profusely branched, 30–120 cm tall
- leaves simple, alternate or fascicled, glandular
- flowers (florets) borne in numerous small heads
- disk florets only, numerous, yellow
- dry open sites, especially alkaline areas in Southwest from Texas to California

Isocoma pluriflora

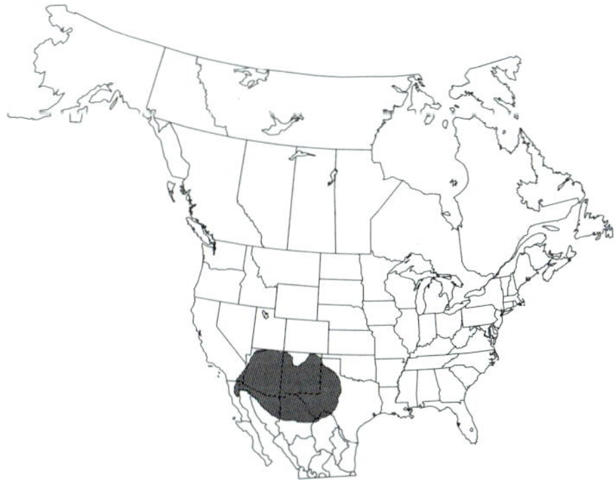

I. pluriflora

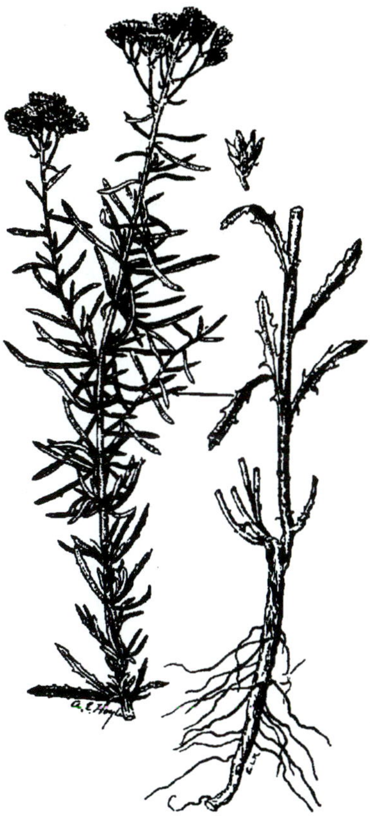

I. pluriflora

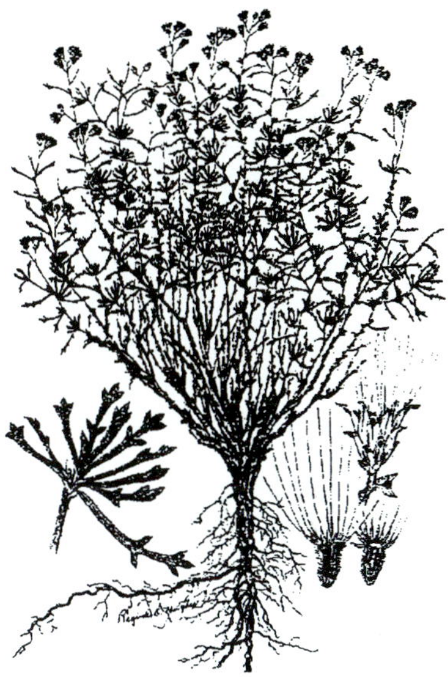

I. tenuisecta

I. tenuisecta

XIII. Tetany and Generalized Tremors: early excitement, later depression, excess salivation, labored respiration, weakness, ataxia, stiffness, muscle tremors and twitching, third-eyelid protrusion

Nicotiana
tobacco
Solanaceae, nightshade family

Signs
- early excitement
- later depression, excess salivation, rapid and labored respiration, ataxia, tremors, severe weakness, muscle twitching or trembling over shoulders and then entire body, peculiar stiff-legged walk, protrusion of third eyelid
- signs last 5–6 hours, survival beyond 12 hours good prognosis
- death from respiratory failure

Pathology
- no distinctive lesions
- analysis of urine or tissues for alkaloids

Treatment
- induce emesis or perform gastric lavage
- administer activated charcoal orally

Disease Problem
- cultivated tobacco risk for pets and pigs
- wild tobaccos seldom eaten
- 0.5% b.w. horses and 2% in cattle depending on plant species
- pyridine alkaloids; nicotine, nornicotine
- sympathetic and parasympathetic ganglionic depolarizers; low dosage stimulation; high dosage blockade
- neuromuscular and CNS effects
- also affects reproductive and urinary systems

Taxonomy and Ecology
- representative species of toxicologic concern:

N. attenuata	coyote tobacco
N. glauca	tree tobacco, mustard tree, buena moza, Don Juan, gigante
N. tabacum	cultivated tobacco
N. trigonophylla	desert tobacco, tabaquillo

- erect aromatic herbs, shrubs, or small trees
- leaves simple, alternate, lower often quite large
- flowers in terminal racemes or panicles, showy, corollas 5-lobed, funnelform, white to yellowish green
- fruits capsules
- cultivated; native species occupy a variety of habitats, some weedy invaders and in dry open areas to thickets

Nicotiana tabacum

N. glauca

N. attenuata

N. glauca

7
Plants Affecting the Reproductive System

Of the various ways that toxic compounds produced by plants alter the animal's organs or systems, perhaps the most cryptic are their effects on the reproductive system. The animal ingesting these plants typically exhibits no or few clinical signs, and it is not until much later at birth of the offspring or during their early development that signs of a toxicologic problem appear. In addition, differentiating between a plant-caused and a gene- or environment-caused reproductive problem is sometimes tenuous, especially if the animal's feeding history and/or the plant composition of its pasture or range is not known. In most instances, only those reproductive problems involving mammary glands and lactation produce clinical signs that can be correlated, more or less immediately, with ingestion of a toxic plant. Nine genera in seven families produce toxicants that specifically affect the reproductive system. An additional 12 genera in eight families produce significant changes in the system in addition to their primary effects on other systems.

Principal Signs

I. Agalactia
II. Mastitis
III. Milk Color Red
IV. Milk Taste Disagreeable
V. Impaired Reproduction
VI. Neonate Small and Weak or Stillborn or Aborted
VII. Teratogenic Skeletal Effects

Plants Causing Problems

I. Agalactia

Primary

Festuca	fescue

Secondary	
Trifolium	clover (page 76)
Tertiary	
Poaceae	grasses; ergotism

II. Mastitis

Primary	
Persea	avocado

III. Milk Color Red

Tertiary	
Caltha	marsh marigold
Ranunculus	buttercup

IV. Milk Taste Disagreeable

Secondary	
Allium	onion, garlic (page 12)
Brassicaceae	mustards (page 14)
Conium	poison hemlock (page 204)
Helenium	sneezeweed (page 66)
Nicotiana	tobacco (page 210)
Ranunculus	buttercup (page 56)
Tertiary	
Anthemis	dog fennel
Hyoscyamus	henbane
Triticum	wheat

V. Impaired Reproduction

Secondary	
Astragalus	locoweed (page 128, 220)
Brassica	mustard (page 14)
Festuca	fescue (page 84, 216)
Gossypium	cottonseed (page 32)
Gutierrezia	broomweed (page 222)
Trifolium	clover (page 76)

VI. Neonate Small and Weak or Stillborn or Aborted

Primary	
Astragalus	locoweed
Gutierrezia	broomweed
Iva	sumpweed

Juniperus	juniper
Pinus	pine
Tertiary	
Cupressus	cypress
Indigofera	wild indigo

VII. Teratogenic Skeletal Effects

Primary	
Lupinus	lupine
Veratrum	false hellebore
Secondary	
Astragalus	locoweed (page 128, 220)
Conium	poison hemlock (page 204)
Nicotiana	tobacco (page 210)
Sorghum	sorghum (page 18, 190)
Tertiary	
Baptisia	wild indigo
Laburnum	golden-chain tree
Thermopsis	golden banner

I. Agalactia: mainly horses; less commonly prolonged gestation, large fetus stillborn and/or dystocia, thickened placenta

Festuca
fescue
Poaceae, grass family

Signs and Diagnostic Aids

- mainly horses
- decrease in milk production or complete lack of udder development
- decreased conception rate
- stillbirths; abortion; firm, thickened placentas
- prolonged gestation results in large fetus, difficult delivery, increased risks to fetus and dam
- foals: weak, overgrown hooves, irregular incisor eruption, long haircoats, poor muscling
- cattle, less in sheep: decreased reproductive efficiency, conception delay, decreased birth weights
- submit plant samples to confirm presence of endophyte via staining and/or ELISA

Pathology

- horses: placenta thick, tough
- placenta microscopically: edema, colloid deposition, vascular degeneration, fibrosis

Treatment and Prevention

- remove horses from fescue forage at least 45–60 days prior to expected delivery date
- horses: perphenazine 0.375 mg/kg b.w. orally 2 times per day for last 30 days of gestation prevents reproductive effects; also dromperidone and fluphenazine

Disease Problem

- ergot-type alkaloids; endophytic fungus, *Neotyphodium coenophialum*, entirely within stems, leaf sheaths, spikelets; does not alter appearance of plant
- infestation levels as low as 20% may be a risk in horses
- ergopeptides, and ergoclavine alkaloids, ergovaline; highest in caryopses/seeds, leaf sheaths (loss of toxicity after 2 years)
- the effects are not related to environmental temperature
- decrease in prolactin secretion and serum levels
- also affects the digestive system
- similar syndrome possible with other forage infected with endophytes; *Lolium, Cyperus*

Taxonomy and Ecology

- three cultivated and lawn species of toxicologic concern:

F. arundinacea	tall fescue, reed fescue
F. pratensis	meadow fescue, English bluegrass
F. rubra	chewings fescue, red fescue

- perennial bunchgrasses with membranous ligules, 50–150 cm tall
- inflorescences panicles
- spikelets laterally compressed, florets 2–35, glumes shorter than lemmas

Festuca arundinacea

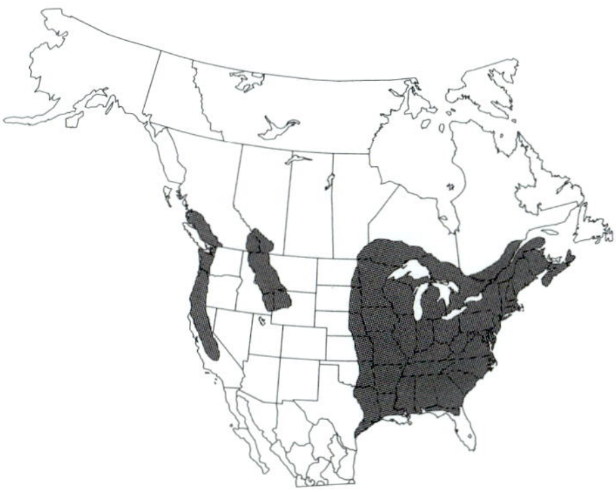

F. arundinacea

Festuca spikelets

F. arundinacea

II. Mastitis: abrupt onset; swollen udder, small amounts of watery milk and/or subcutaneous edema, respiratory distress, cardiac failure

Persea
avocado
Lauraceae, laurel family

Signs and Diagnostic Aids

- udder hard, swollen, 75% decreased milk production
- milk watery, cheesy, curdled
- subcutaneous edema of neck and brisket, infrequent cough, weakness/depression, reluctance to move, respiratory distress, cardiac arrhythmias
- increase in serum LDH, CPK, AST
- horses: severe edema of head, neck, ventral areas
- birds: listless, respiration labored

Pathology

- mammary glands: edematous, reddened, clots in large ducts
- edema prominent in many tissues: subcutis, lungs, gall bladder, perirenal tissues
- free fluid: heart sac, thorax, abdomen
- heart: pale, flabby
- microscopically: edema, necrosis of secretory acinar epithelium of mammary gland; interstitial edema and myocardial degeneration of heart, necrosis especially ventricular walls and septum

Treatment

- provide NSAIDs and diuretics
- administer antibiotics for secondary infection in mammary gland

Disease Problem

- confirmed for only Guatemalan-type avocado and hybrids, Fuerte
- fruits, seeds, and especially leaves
- goats particularly sensitive
- low dosage (2% b.w.): mastitis in lactating animals
- higher dosage (3% b.w.) or repeated low dosage and in nonlactating animals: cardiac effects, especially horses
- canaries, budgerigars, cockatiels, ostriches: subcutaneous edema of neck, chest, and pulmonary system
- persin, a long-chain, unsaturated compound

Taxonomy and Ecology

- one species of toxicologic concern:

  P. americana avocado, alligator pear, aguacate

- shrubs or trees, dense crown
- leaves alternate; single, primary vein
- fruits berries (avocado); single, large seed
- three ecological races, numerous cultivars; skin of fruits smooth, thin for Mexican, warty for Guatemalan, leathery for West Indian; Fuerte, a hybrid between Guatemalan and Mexican, most popular cultivar; cultivated extensively in Mexico, Florida, California

Persea americana

P. americana

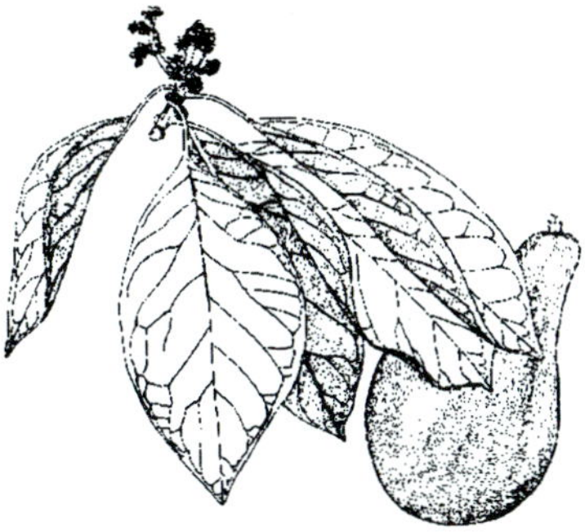

P. americana

VI. Neonate Small and Weak or Stillborn or Aborted: also fetal deformities of joints, thoracic limbs, tendon contraction

Astragalus
locoweed
Fabaceae, pea family

Signs

- small and weak neonates, unable or lack desire to suckle
- abortions
- less commonly fetal deformities: thoracic limb rotation, arthrogryposis, tendon contraction
- adult male: decrease in libido, sperm motility and viability
- adult female: degeneration of ovarian luteal cells, significant decrease in serum progesterone, overall decrease in fertility
- may be accompanied by signs of locoism in some mature animals

Pathology

- microscopically: cytopathic vacuolation of cells of various tissues

Treatment and Prevention

- avoid pasturing animals in spring and early summer where significant populations of locoweed are present

Disease Problem

- last trimester of gestation for abortions and weak neonate
- ingestion of 25–30% b.w. of plant over 3–4 weeks
- presumably due to swainsonine and accumulation of oligosaccharides compromising cellular function
- also affects cardiac and nervous systems

Taxonomy and Ecology

- representative species of toxicologic concern:

A. lentiginosus	speckled loco, freckled milk vetch
A. mollissimus	woolly loco, purple loco
A. wootonii	garboncillo, Wooton's locoweed

- annual or perennial herbs, low growing, decumbent to erect
- leaves compound, terminal leaflets and stipules present
- flowers pealike, petals 5, white, pink, violet
- legumes sometimes inflated
- almost exclusively in western half of continent; in grasslands, deserts, alpine tundra, open woods

Astragalus mollissimus

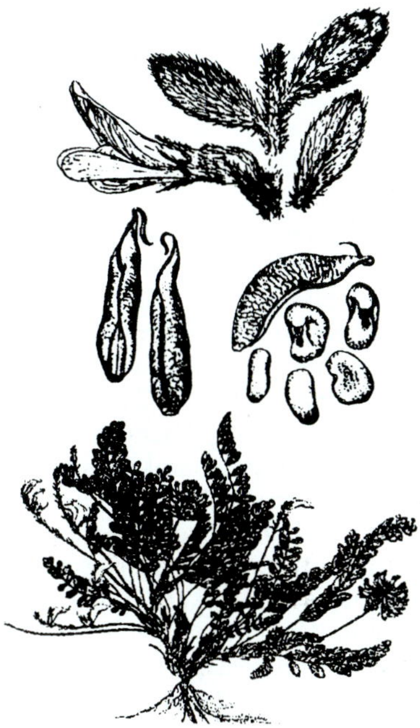

A. mollissimus

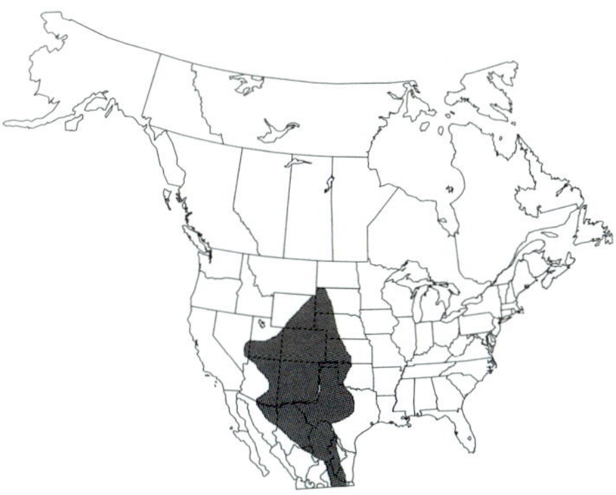

A. mollissimus

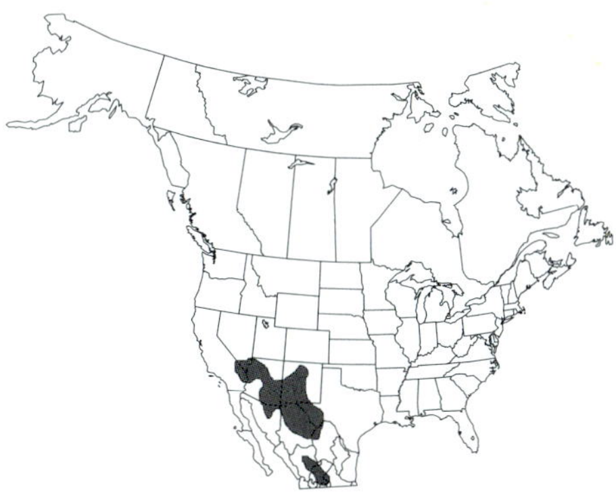

A. wootonii

A. lentiginosus

VI. Neonate Small and Weak or Stillborn or Aborted: cattle, sheep, goats; premature parturition; adult may be ill; placental retention

Gutierrezia
broomweed
Asteraceae, sunflower family

Signs and Diagnostic Aids

- vulvar swelling, mammary development several weeks or months prior to expected parturition
- calves: abortion, stillbirth, or premature, small, weak (12–20 kg b.w.); some survive less than 48 hours
- retention of placental membranes
- adult: crusty nasal pad, mucopurulent nasal discharge; decreased appetite, weakness, listlessness; stringy mucoid secretion from vulva; diarrhea and/or constipation
- increase in creatinine and bilirubin (both direct and indirect) in adult

Treatment

- provide general nursing care
- provide supplemental feeding
- progesterone or vitamin A not preventive

Disease Problem

- sheep, cattle at greatest risk, goats somewhat less so
- plants growing on sandy soils primary risk, especially in late winter or early spring
- 2.5–5% b.w. eaten over 1–2 weeks may present a risk
- with heavy plant infestations, calf crop down to 60%
- toxin unknown, possibly diterpene acids as in *Pinus*

Taxonomy and Ecology

- two perennial species of toxicologic concern:

G. microcephala	threadleaf snakeweed, small-headed matchweed, small-headed matchbrush
G. sarothrae	snakeweed, broom snakeweed, matchweed, matchbrush, turpentine weed, coyaye, perennial broomweed, slinkweed, turpentine weed, hierba de vibora

- subshrubs or shrubs, 10–60 cm tall
- leaves simple, alternate
- foliage, bright green and aromatic in early season
- flowers (florets) borne in numerous small heads
- ray florets 1–7; yellow
- disk florets 1–13; yellow
- variety of habitats; western half of continent

Gutierrezia sarothrae

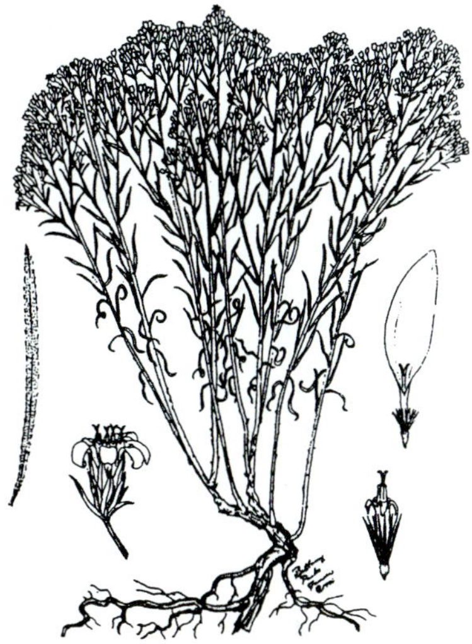

G. sarothrae

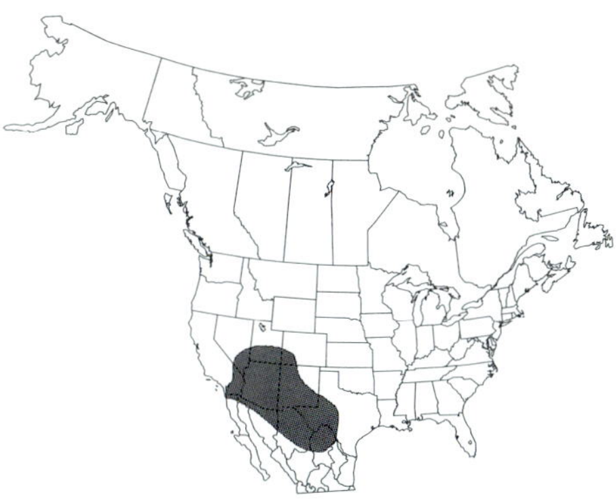

G. microcephala

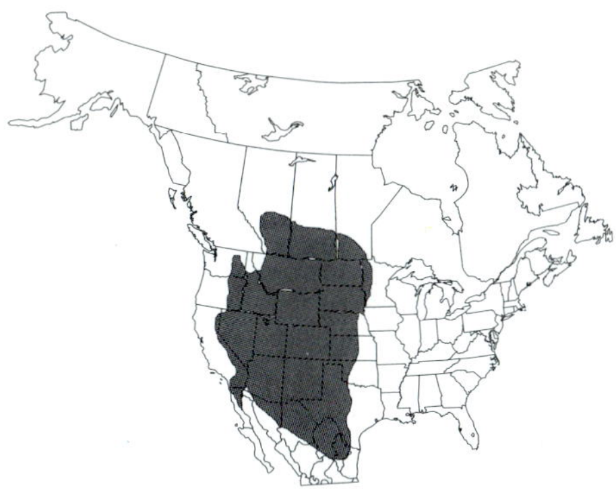

G. sarothrae

VI. Neonate Small and Weak or Stillborn or Aborted: cattle; premature parturition

Iva
sumpweed
Asteraceae, sunflower family

Signs
- premature mammary development, dripping milk, excess vaginal mucous
- calves: stillbirth, or premature and/or weak animals, possibly abortions
- poor survival of newborn

Pathology
- placental mineralization but not retention

Treatment and Prevention
- delay breeding to avoid ingestion of seedlings during last half of gestation

Disease Problem
- toxic in all growth stages but mainly seedlings in 2–8 leaf stage
- may be quite abundant and unavoidably grazed
- ingestion for several weeks
- other species of genus not similarly toxic
- animal susceptible during last half of gestation
- toxicant unknown

Taxonomy and Ecology
- one species of known toxicologic concern:

I. angustifolia	narrowleaf sumpweed

- erect, annual herb, 50–120 cm tall
- leaves simple, alternate, linear
- flowers (florets) borne in numerous small, axillary heads
- disk florets only, 5–7, staminate or pistillate, translucent
- well-drained upland areas or along ditches, creek beds, moist disturbed sites; south-central part of continent

Iva angustifolia

I. angustifolia

I. angustifolia

VI. Neonate Small and Weak or Stillborn or Aborted: cattle; premature parturition; adult may be ill

Juniperus
juniper
Cupressaceae, cypress family

Signs
- calves: premature small, weak, or stillborn
- dam may appear ill

Pathology
- no specific lesions

Treatment and Prevention
- prevent access to trees during late pregnancy

Disease Problem
- *J. communis* especially toxic
- mainly cattle; less than 1% b.w. over several days
- sheep only with exceptionally high dosage, 5–6 times as much
- foliage is commonly eaten in small amounts by deer
- possibly all species of genus toxic but confirmed in only a few
- diterpene acids, isocupressic acid, as in *Pinus*

Taxonomy and Ecology
- representative species of toxicologic concern:

J. communis	common juniper, little juniper, dwarf juniper
J. deppeana	alligator juniper
J. horizontalis	creeping juniper, trailing juniper
J. occidentalis	western juniper, Sierra juniper
J. osteosperma	Utah juniper, little Utah juniper
J. sabina	savin
J. scopulorum	Rocky Mountain juniper, red cedar
J. virginiana	eastern red cedar, Virginia juniper

- evergreen shrubs or trees, strongly aromatic, most dioecious, 3–30 m tall
- leaves simple, opposite in 4 ranks or whorled
- pollen cones yellow
- seed cones fleshy, berrylike
- variety of vegetation types and habitats across continent

Juniperus virginiana

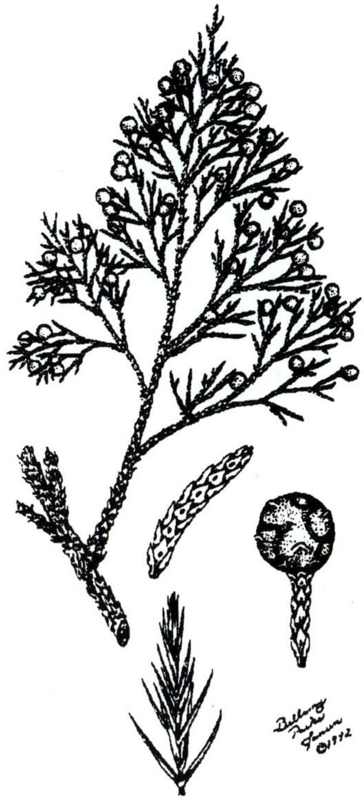

Juniperus

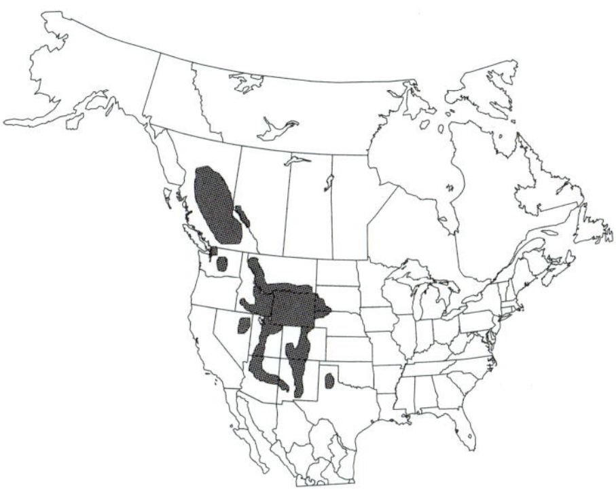

J. scopulorum

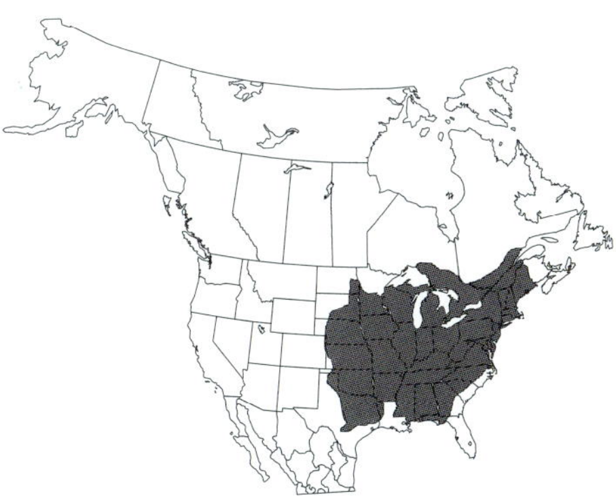

J. virginiana

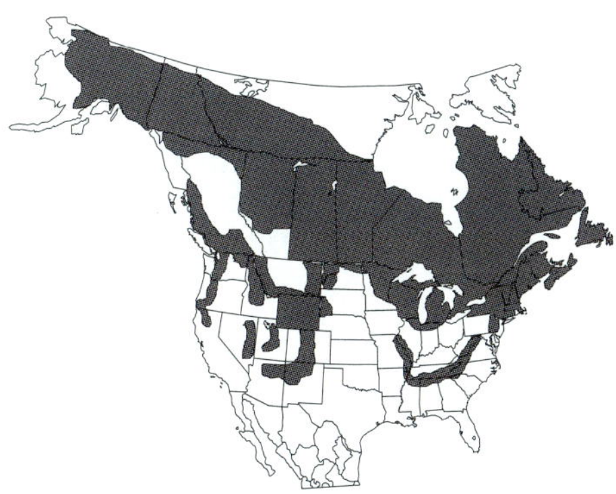

J. communis

VI. Neonate Small and Weak or Stillborn or Aborted: cattle; premature parturion; adult may be ill

Pinus
pine
Pinaceae, pine family

Signs
- last trimester of pregnancy: swelling of vulva, mammary development, blood-tinged mucoid vulvar discharge
- calves: abortion or stillbirth or premature, weak newborn; poor survival
- some cows: little mammary development or milk letdown
- cows: loss of appetite, lack of ruminal activity, depression, weakness
- retention of fetal membranes, metritis, uterus flaccid, doughy, fluid-filled

Pathology
- endometritis with ulceration

Treatment
- remove placenta
- treat for metritis

Disease Problem
- needles readily eaten; small amounts (< 0.5 kg) not a problem
- needles, green or dry, >1 kg, mainly last 2 months of pregnancy; consumption increases with low temperature and snow cover
- effects seen days to weeks after ingestion
- only cattle, bison
- sheep, deer little affected
- diterpene acids, isocupressic acid; impairment of increase in uterine blood flow needed for growing fetus, decreased viability of fetus
- other species of genus also may be toxic

Taxonomy and Ecology
- four species of known toxicologic risk:

P. contorta	lodgepole pine, limber pine
P. jeffreyi	Jeffrey pine
P. ponderosa	ponderosa pine, western yellow pine, Arizona pine, Arizona ponderosa pine, Pacific ponderosa pine
P. radiata	Monterey pine

- evergreen trees, resinous, deeply furrowed bark
- leaves (needles) in bundles of 2–5, rarely 1 or 6
- pollen cones, small, soft
- seed cones mature in 2–3 years
- variety of sites; mountains; western half of continent

Pinus ponderosa

P. ponderosa

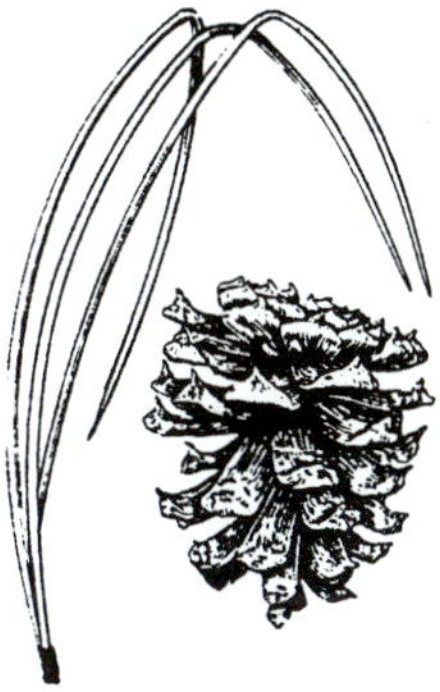

P. ponderosa

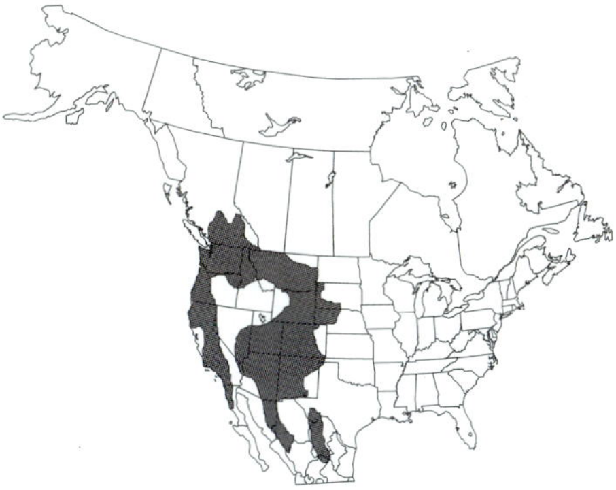

P. ponderosa

VII. Teratogenic Skeletal Effects: cattle; deformities of joints, long bones, vertebrae, palate

Lupinus
lupine
Fabaceae, pea family

Signs

- calves: arthrogryposis of limbs; elbow joints immobile, flexed, unable to be extended due to misalignment; malpositioning of ulnar-humeral articulation, lateral rotation distal to elbow
- 10–15% with limb, vertebral, palate defects
- unable to follow dam and nurse; many die
- additional deformities: lordosis, scoliosis, torticolis, cleft palate

Treatment

- if defects not severe, calf may outgrow them or compensate

Disease Problem

- calves and less commonly goats
- "crooked calf disease"
- days 40–70 of gestation critical time period
- eaten continuously for several weeks
- deformation may be due to in utero depression of fetal movement rather than or accompanying outright malformation
- quinolizidine alkaloids, anagyrine and baptifoline, piperidine alkaloids, ammodendrine and N-acetylhystrine, depress fetal activity
- possible public health risk due to alkaloids in milk
- also affects digestive and neuromuscular systems

Taxonomy and Ecology

- representative species known to cause reproductive effects:

L. arbustus	spur lupine
L. argenteus	silvery lupine, tailcup lupine
L. formosus	grassland lupine
L. latifolius	broad-leaved lupine
L. leucophyllus	woolly-leaved lupine, western lupine, velvet lupine, white-leaved lupine
L. sericeus	silky lupine, sink lupine
L. sulphureus	sulfur lupine

- erect herbs or subshrubs, up to 1 m tall
- leaves palmately compound, alternate, leaflets 3–17, typically 5
- inflorescences terminal racemes or spikes
- flowers pealike, petals 5, blue, violet, lilac, crimson, white, or yellow
- legumes flattened, seeds 2–12
- wide range of habitats; grasslands, shrublands, open forests; greatest abundance in western half of continent

Lupinus latifolius

L. argenteus

L. leucophyllus

L. leucophyllus

VII. Teratogenic Skeletal Effects: sheep mainly; monkey-faced lambs sometimes with a single eye; deformities of joints, long bones, and trachea

Veratrum
false hellebore
Liliaceae, lily family

Signs

- lambs: born with single or double globe, reduced or nonexistent eye; proboscis above or below globe or absent
- shortened maxilla/premaxilla, upward curvature of the mandible
- cebocephalus (short snout); limited cerebral development; single optic nerve; prolonged gestation, due to absence of pituitary gland
- sometimes extralarge stillbirth
- other deformities: bilateral shortening of metacarpals, metatarsals, and tibias; arthrogryposis of fetlocks, carpals, hocks, and stifles; deformities or fusing of articulating surfaces; lateral flattening of trachea obliterating lumen
- less common in calves
- adult intoxication: abrupt onset; excess salivation with froth around mouth, frequent swallowing, vomiting
- fetal effects may occur without overt signs of illness in dam

Pathology

- adult: no significant lesions

Treatment and Prevention

- restrict access to plants during first trimester of gestation
- administer atropine for intoxication in adult

Disease Problems

- monkey-faced lambs; cyclopia, other types of cephalic malformations; consumption of plants on day 14 of gestation
- little risk of deformities after sixth week of gestation for sheep, pigs; after eighth week for cattle, horses
- most animal species are susceptible
- nitrogenous steroidal alkaloids, azasteroids; jervanines, cyclopamine most important
- also affects cardiovascular, digestive and neuromuscular systems

Taxonomy and Ecology

- first species of known teratogenic risk; others suspect:

V. californicum	California false hellebore, corn-lily, skunk cabbage
V. insolitum	Siskiyou false hellebore
V. parviflorum	hellebore
V. viride	American white hellebore, green false hellebore, skunk cabbage
V. woodii	Indian-poke, Wood's false hellebore

- robust, erect herbs, from thick, black rhizomes, 1–2 m tall
- leaves alternate, numerous, some clasping
- inflorescences in terminal panicles
- flowers white, greenish, yellowish-green, reddish-brown, or purple; perianth parts 6; stamens 6
- moist, open sites in alpine meadows, open woodlands, marshes, along waterways, lake edges in mountains

Veratrum californicum

V. californicum

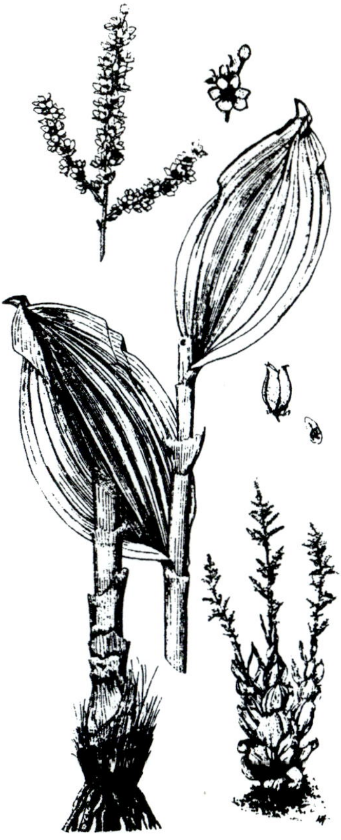

V. californicum

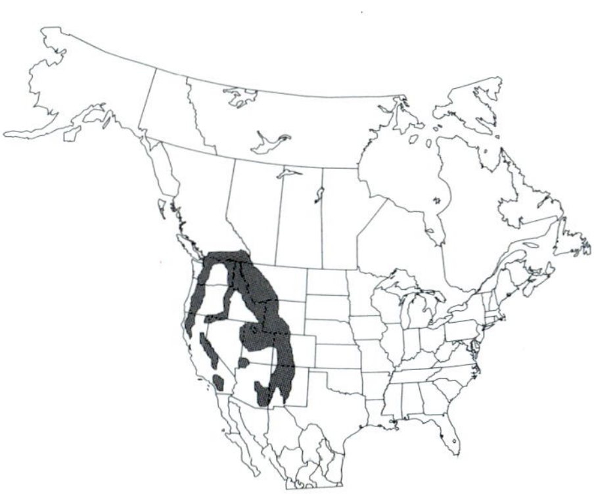

V. californicum

8
Plants Affecting the Respiratory System

In many cases of plant-caused intoxication, the respiratory system may exhibit changes, such as increases or decreases in respiratory rate, edema in the lungs, or labored respiration, as the ingested toxicants act on other organs and/or systems. Only two genera in two families are the primary cause of significant changes solely in the respiratory system. An additional 12 genera produce changes in addition to their primary effects on other systems.

Principal Signs

I. Expired Air with Distinct Odor
II. Respiratory Distress Moderate
III. Respiratory Distress Severe (ARDS)

Plants Causing Problems

I. Expired Air with Distinct Odor

Secondary

Allium	onion, garlic, leek (page 12)
Eupatorium	white snakeroot (page 206)
Isocoma	goldenbush (page 208)

II. Respiratory Distress Moderate

Secondary

Asclepias	milkweed (page 24, 164)
Eupatorium	white snakeroot (page 206)
Gossypium	cottonseed (page 32)
Helenium	sneezeweed (page 66)
Hymenoxys	bitterweed (page 68)
Melia	Chinaberry (page 182)

Tertiary	
Flourensia	whitebrush
Lobelia	lobelia

III. Respiratory Distress Severe (ARDS)

Primary	
Medicago	alfalfa
Perilla	purple mint
Secondary	
Brassica	mustard (page 14)
Cynodon	Bermudagrass (page 196)
Panicum	panicum (page 260)
Tertiary	
Ipomoea	sweet potato
Pennisetum	buffelgrass
Phaseolus	scarlet runner bean
Poaceae	grasses
Sphenosciadium	white-heads

III. Respiratory Distress Severe: abrupt onset; mainly cattle; labored open-mouth respiration, foamy salivation; ARDS

Medicago
alfalfa
Fabaceae, pea family

Signs

- abrupt onset; labored respiration, loud expiratory grunts, open-mouth breathing; frothy, excessive salivation
- subcutaneous emphysema of neck, shoulders, back
- death often 1–2 days after onset
- ARDS mainly in cattle but other ruminants at risk

Pathology

- lungs fail to collapse; heavy, firm, froth-filled airways; interlobular septae distended with fluid, fibrin, and air bubbles
- cut surfaces of lung: marbled appearance with patchy areas of firm, reddened to purple lobules, prominent edema in connective tissue and septa, and air bubbles
- lungs microscopically: alveolar epithelialization, accumulations of alveolar macrophages, hyaline membrane formation, proliferation of type II pneumocytes

Treatment and Prevention

- poor response to treatment, avoid stress
- antiprostaglandin drugs, flunixin meglumine may help
- administer atropine
- prevent by gradual changes in forage quality; cut lush pasture and allow to wilt, limit grazing time on new pasture, rotate animals between the lush and dry
- supplement with an ionophore, monensin (200 mg/animal) 10 days before and after forage change

Disease Problem

- "ARDS" (acute respiratory distress syndrome); also known as "acute bovine pulmonary emphysema"
- mainly mature cattle; appearance 5–10 days after an abrupt change in diet from poor to lush green forage
- sudden ingestion of high levels of L-tryptophan; bacterial conversion in rumen to 3-methylindole (3-MI); absorbed and subsequently metabolized in lungs to compound(s) toxic to pneumocytes, leading to pulmonary emphysema and edema
- morbidity rate usually low, but rarely may exceed 50%; case fatality rate 25–50%

Taxonomy and Ecology

- one cultivated species of toxicologic concern:

  M. sativa alfalfa, lucerne

- deep-rooted, perennial herb, 50–100 cm tall
- leaves 1-pinnately compound, leaflets 3, margins dentate
- flowers small, pealike, blue to purple
- introduced; cultivated for forage

Medicago sativa

M. sativa

M. sativa

III. Respiratory Distress Severe: abrupt onset, labored open-mouth respiration, foamy salivation; ARDS

Perilla
purple mint
Lamiaceae, mint family

Signs

- abrupt onset; labored respiration, loud expiratory grunts, open-mouth breathing; frothy, excessive salivation
- subcutaneous emphysema of neck, shoulders, back; death often 1–2 days after onset

Pathology

- lungs fail to collapse; heavy, firm, froth-filled airways; interlobular septae distended with fluid, fibrin, and air bubbles
- cut surfaces of lung: marbled appearance of patchy areas of firm, reddened to purple lobules, prominent edema in connective tissue and septa, and air bubbles
- lungs microscopically: alveolar epithelialization, accumulations of alveolar macrophages, hyaline membrane formation, proliferation of type II pneumocytes

Treatment

- poorly responsive to treatment
- avoid stress
- administer atropine as needed

Disease Problem

- "ARDS" (acute respiratory distress syndrome); also known as "panting disease"
- in contrast to typical ARDS, dyspnea from *Perilla* may occur in horses and other animal species as well as ruminants
- mainly late summer and early fall, flowering and fruiting plants
- 2–3% b.w. of plants eaten for 1–3 days
- volatile, aromatic 3-substituted furan ketones; short-lived intermediates damage type I pneumocytes; excessive proliferation of type II cells leads to glandular appearance of lung tissue in 12–24 hours

Taxonomy and Ecology

- one introduced species of toxicologic concern:

P. frutescens	purple mint, perilla, perilla mint, beefsteak plant, wild coleus, Joseph's coat, rattlesnake weed, blue weed

- strongly aromatic, erect, annual herb, 20–175 cm tall
- stems square or 4-sided
- leaves simple, opposite, typically purplish
- inflorescences elongate racemes
- flowers bilabiate, petals 5, purple, lavender, or white
- naturalized; dense stands, semishaded sites; southeastern quarter of continent

Perilla frutescens

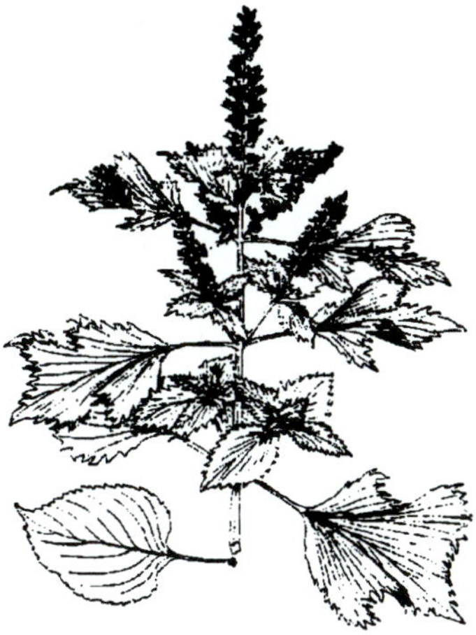

P. frutescens

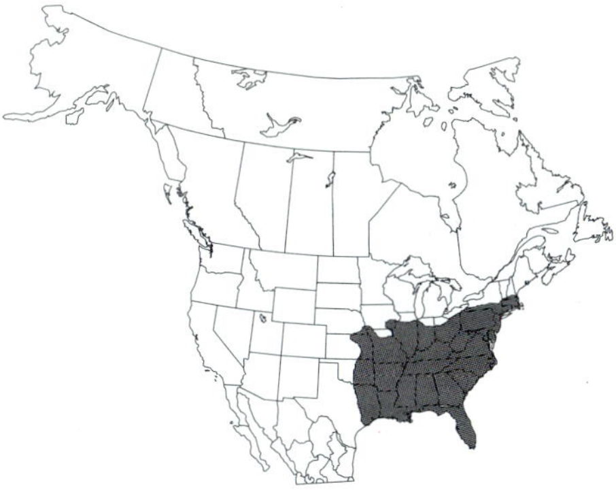

P. frutescens

9
Plants Affecting the Skin

The skin of an animal may be affected by toxic plants either directly via mechanical contact, as happens when animals brush against plants with stinging hairs, or indirectly by the ingestion of compounds that either directly or indirectly damage the skin. Only two genera in different families are of concern with respect to mechanical irritation of the skin. In contrast, eight genera in seven families produce compounds that induce photosensitization, a condition of increased sensitivity to ultraviolet light and the formation of transient reactive radicals in the peripheral circulation that affect adjacent tissues and thus manifest skin changes. An additional 13 species produce alterations in the skin in addition to their primary effects on other organs or systems.

Principal Signs

I. **Hair Loss**
II. **Irritant Effects:** reddening, blisters, pruritis
III. **Necrosis and Sloughing:** pigmentation pattern dependent; primary (direct) photosensitization
IV. **Necrosis and Sloughing:** pigmentation pattern dependent; secondary (hepatogenous) photosensitization
V. **Necrosis and Sloughing:** pigmentation pattern independent
VI. **Necrosis of Distal Limbs, Ears, or Tail, with Lameness**

Plants Causing Problems

I. Hair Loss

Secondary

Astragalus	locoweed (page 128, 220)
Crotalaria	rattlebox (page 106)
Prosopis	mesquite (page 156)
Vicia	vetch (page 264)

Tertiary

Leucaena	lead tree

II. Irritant Effects: reddening, blisters, pruritis

Primary	
Cnidoscolus	bull nettle
Urtica	stinging nettle
Tertiary	
Daphne	daphne
Dirca	leatherwood
Euphorbia	spurge
Hippomane	manchineel
Hura	sandbox tree
Laportea	wood nettle
Primula	primrose
Tragia	noseburn

III. Necrosis and Sloughing: pigmentation pattern dependent; primary (direct) photosensitization

Primary	
Ammi	bishop's weed
Hypericum	St. John's wort
Thamnosma	blisterweed
Secondary	
Brassica	mustard (page 14)
Cynodon	Bermudagrass (page 196)
Trifolium	clover (page 76)
Tertiary	
Astragalus cicer	chickpea milk vetch
Avena	oats
Cenchrus	sandbur
Colubrina	hog plum
Cooperia	rain lily
Cymopterus	spring parsley
Descurainia	tansy mustard
Echinochloa	barnyard grass
Eriogonum	wild buckwheat
Fagopyrum	buckwheat
Heracleum	hogweed
Hordeum	barley
Medicago	alfalfa
Polygonum	knotweed, suspect
Ruta	rue
Secale	rye
Setaria	bristlegrass
Sorghum	sudan
Sphenosciadium	white-heads

Thlaspi	fanweed
Triticum	wheat

IV. Necrosis and Sloughing: pigmentation pattern dependent; secondary (hepatogenous) photosensitization

Primary	
Agave	lechuguilla
Nolina	sacahuista
Panicum	panicum
Tribulus	goathead
Secondary	
Cynodon	Bermudagrass (page 196)
Kochia	kochia (page 86, 160)
Lantana	lantana (page 100)
Lolium	perennial ryegrass (page 198)
Phyllanthus	leaf-flower (page 102)
Tetradymia	horsebrush (page 104)
Trifolium	clover (page 76)

V. Necrosis and Sloughing: pigmentation pattern independent

Primary	
Vicia	vetch

VI. Necrosis of Distal Limbs, Ears, or Tail, with Lameness

Secondary	
Festuca	fescue (page 84, 216)
Tertiary (ergot in pasture grasses)	
Agrostis	bentgrass
Bromus	bromegrass
Dactylis	orchardgrass
Elymus	wild rye
Koeleria	Junegrass
Poa	bluegrass

II. Irritant Effects: localized erythema, blisters, intense pruritis; dogs, horses

Cnidoscolus
bull nettle
Euphorbiaceae, spurge family

Signs
- erythema, swelling, and pruritis at contact site; lasting 20–45 minutes

Treatment
- generally not required
- apply topical local anesthetic for relief

Disease Problem
- mainly dogs upon incidental contact
- massive exposure unlikely due to low density of plants
- horses: react at contact sites on limbs and abdomen
- rarely cattle
- stinging hairs (trichomes) with multicellular pedestal and single, elongated, tapered, hollow cell with a slightly swollen tip
- tip shears off at oblique angle as hair penetrates skin to release contents
- chemical mediator identity unknown but similar in activity to serotonin

Taxonomy and Ecology
- three species of toxicologic concern:

C. angustidens	mala-mujer
C. stimulosus	bull nettle, finger-root, tread-softly
C. texanus	bull nettle, mala-mujer, spurge nettle

- herbs or shrubs, often from massive rootstalks; viscid white sap; stout stinging hairs, 30–100 cm tall
- leaves simple, alternate, palmate venation
- flowers imperfect, sepals white, petals absent
- deep sandy soils of southern portion of continent

Cnidoscolus texanus

C. texanus

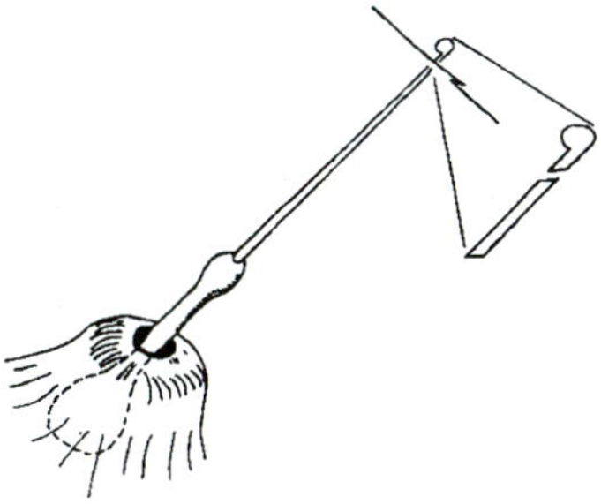

stinging hair

C. texanus

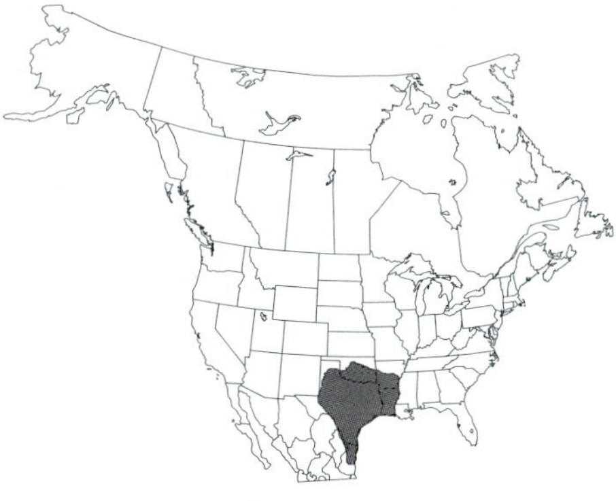

C. texanus

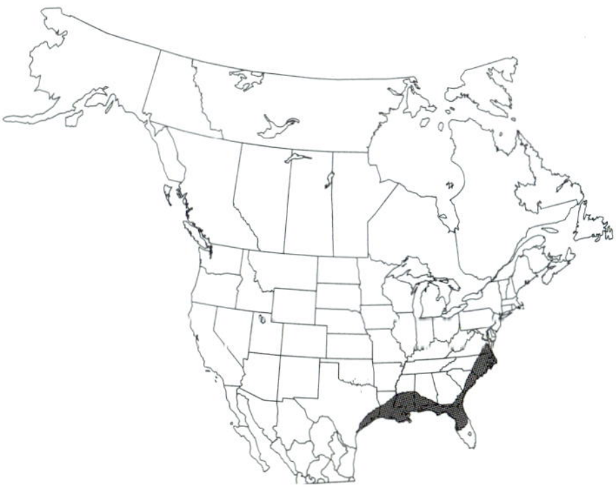

C. stimulosus

II. Irritant Effects: localized or extensive erythema, blisters, intense pruritis; dogs, horses

Urtica
stinging nettle
Urticaceae, nettle family

Signs
- erythema, swelling, pain, pruritis at contact site; lasting minutes to hours
- dogs: more severe; within minutes or up to hour after exposure, begin salivating profusely, licking paws, scratching at nose, face, and other body areas, retching, and vomiting
- dogs: massive or repeated contact; rapid and/or labored respiration, vomiting, hemorrhage from nostrils, muscle fasciculations, ataxia to partial paresis of pelvic limbs; signs subside within hours or days

Treatment
- administer atropine and antihistamines
- provide supportive care

Disease Problem
- mainly dogs, especially hunting breeds exposed to many plants in open woods or bottomlands; late winter/early spring
- horses: react at contact sites on limbs and abdomen
- rarely cattle; cattle, sheep, rabbits may eat plants as forage
- stinging hairs (trichomes) with slender, hollow shaft arising from bulbous pedestal
- bulbous tip shears off at oblique angle as hair penetrates skin to release contents
- toxicants unknown; histaminic, acetylcholinergic, serotinergic compounds suspected

Taxonomy and Ecology
- four species of toxicologic concern:

U. chamaedryoides	heart-leaf nettle, weak nettle
U. dioica	American stinging nettle, tall nettle, slim nettle, hoary nettle
U. gracilenta	mountain nettle
U. urens	dwarf nettle, dog nettle, burning nettle

- erect annual or perennial herbs armed with stinging hairs, 0.50–2.0 m tall
- leaves simple, opposite
- flowers tiny, axillary
- moist open woodlands, bottomlands, waste areas; across continent

Urtica dioica

U. dioica

U. dioica

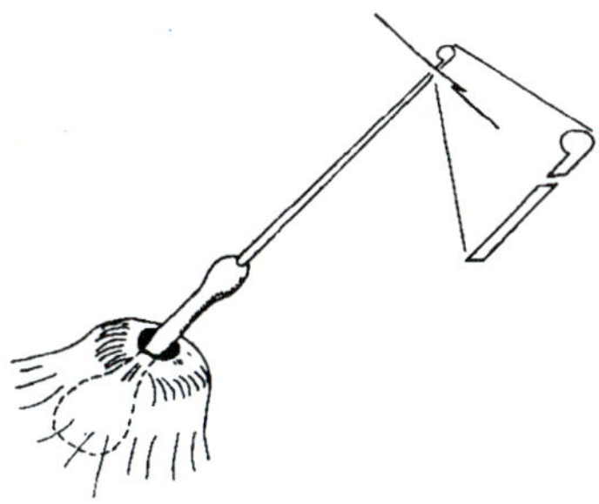

stinging hair

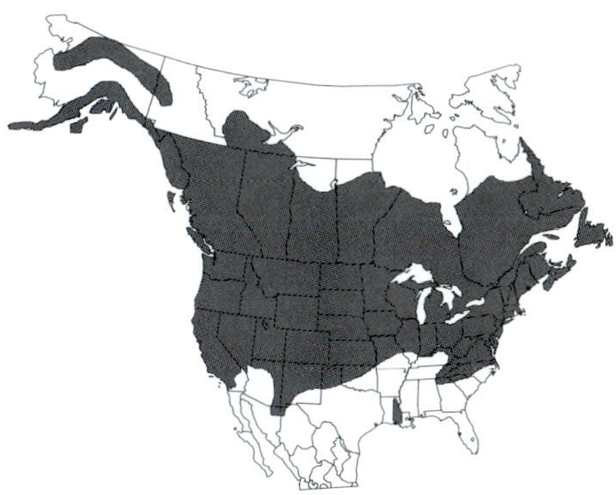

U. dioica

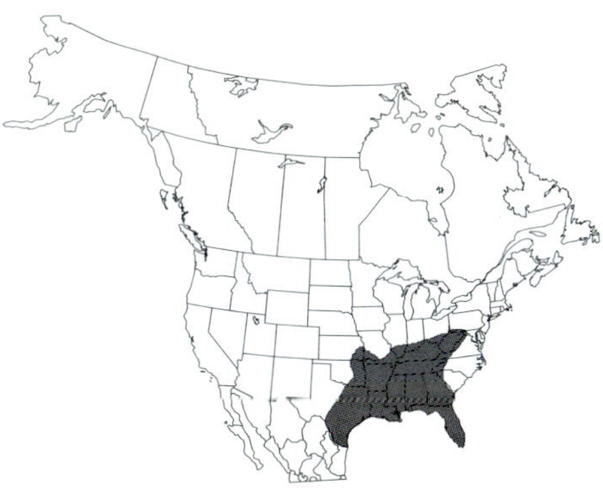

U. chamaedryoides

III. Necrosis and Sloughing: pigmentation pattern dependent; primary (direct) photosensitization; pruritis, erythema, swelling and sloughing of light-colored or thinly haired areas

Ammi
bishop's weed
Apiaceae, carrot family

Signs and Diagnostic Aids

- severe pruritis, reddening, swelling, edema, sloughing of skin of teats, udder, escutcheon, less-pigmented and lightly haired areas
- severe cases: skin thickened, gelatinous appearance extending into deeper corium; crusty, ulcerative, or proliferative areas especially around the face
- lesion distribution coincides with lighter-colored areas
- may have diarrhea, restlessness, agitation, loss of appetite, and increased respiratory rate, occasionally violent behavior at disease onset
- sheep: lesions may be limited to ears, eyelids, face, lips, coronets
- signs typically last for 2 weeks, healing in 2 or more months
- eyes: photophobia, cloudy corneas, reddening of edges of cornea and conjunctiva

Pathology

- microscopically: necrosis, polymorphonuclear infiltrate in deeper corium

Treatment

- provide pain relief, soothing skin medications
- feed and water in shade
- nursing offspring may require supplemental feeding

Disease Problem

- mainly cattle, sheep, and poultry
- seeds are especially toxic
- furanocoumarins activated in peripheral circulation by UV light to form transient reactive radicals that cause damage to surrounding tissues
- effects seldom lethal
- possible systemic involvement

Taxonomy and Ecology

- two introduced species of toxicologic concern:

A. majus	bishop's weed, greater ammi
A. visnaga	bisnaga, Bishop's weed, toothpick ammi, khella, chellah

- erect, branching annuals or biennials, 20–80 cm tall
- leaves 3-pinnately compound or dissected
- flowers borne in compound umbels, white
- naturalized; disturbed sites; mainly southeastern portion of continent

Ammi majus

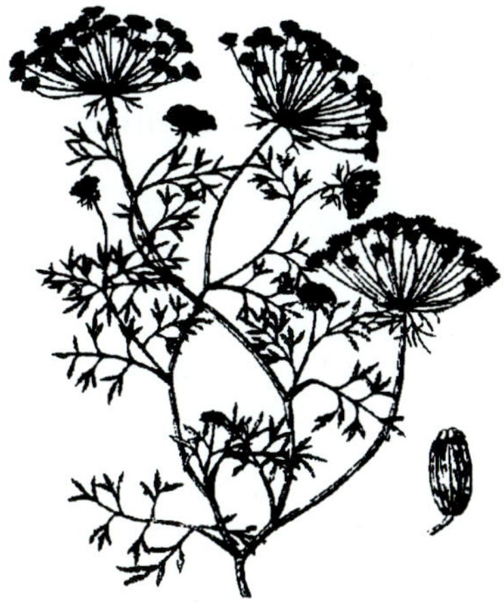

A. majus

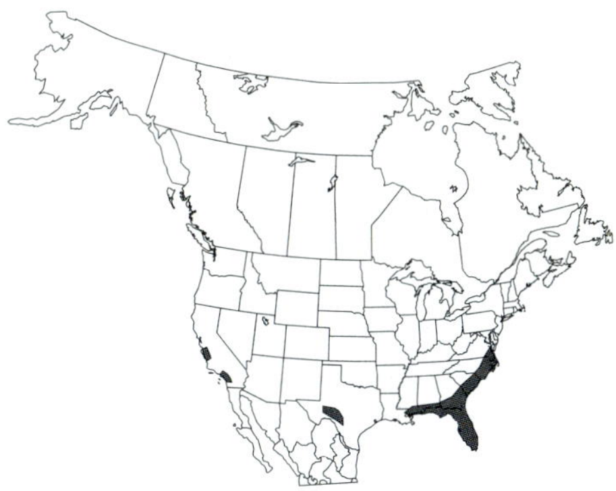

A. visnaga

A. majus

III. Necrosis and Sloughing: pigmentation pattern dependent; primary (direct) photosensitization; pruritis, erythema, swelling and sloughing of light-colored or thinly haired areas

Hypericum
St. John's wort
Clusiaceae, mangosteen family

Signs and Diagnostic Aids

- severe pruritis, reddening, swelling, edema, sloughing of skin of teats, udder, escutcheon, less-pigmented and lightly haired areas
- severe cases: skin thickened, gelatinous appearance extending into deeper corium; crusty, ulcerative, or proliferative areas especially around face
- lesion distribution coincides with lighter-colored areas
- may have diarrhea, restlessness, agitation, loss of appetite, and increased respiratory rate, occasionally violent behavior at disease onset
- sheep: lesions may be limited to ears, eyelids, face, lips, coronets
- signs typically last for 2 weeks, healing in 2 or more months
- eyes: photophobia, cloudy corneas, reddening of edges of cornea and conjunctiva
- sheep: increase in BUN, serum AST, ALT, CK, LDH; decrease in blood glucose, hemoglobin, PCV

Pathology

- microscopically: necrosis, polymorphonuclear infiltrate in deeper corium

Treatment

- provide pain relief, soothing skin medications
- feed and water in shade
- nursing offspring may require supplemental feeding

Disease Problem

- all animal species; however, deer and goats at lower risk
- 1% b.w. eaten over several days may cause problems
- highly condensed quinone hypericin absorbed into the systemic circulation and activated in peripheral circulation by UV light; reacts with lipids, amino acids, and nucleic acids to cause necrosis of adjacent tissues
- effects seldom lethal, but distress affects production of meat, milk, and wool

Taxonomy and Ecology

- representative species of toxicologic concern:

H. concinnum	gold-wire
H. formosum	western St. John's wort
H. perforatum	common St. John's wort, Klamath weed,amber ,goatweed, goatsbeard
H. pseudomaculatum	large spotted St. John's wort
H. punctatum	spotted St. John's wort

- erect herbs or shrubs, 15–120 cm tall
- leaves simple, opposite, or whorled, gland-dotted, glands black or translucent
- flowers showy, petals 5, bright yellow, glands present or absent
- fruits capsules
- native and naturalized; open habitats; across continent

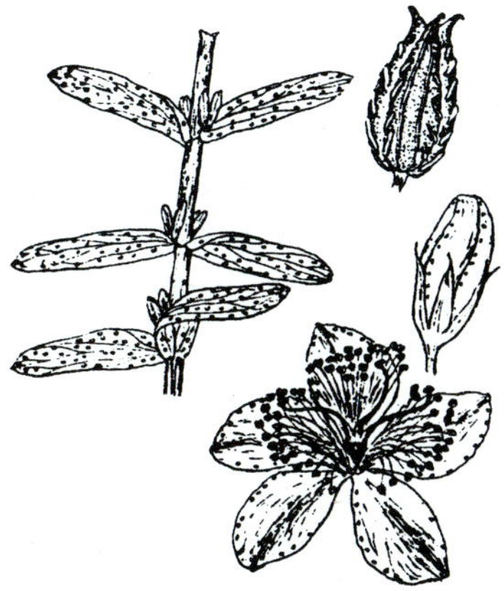

Hypericum perforatum

H. perforatum

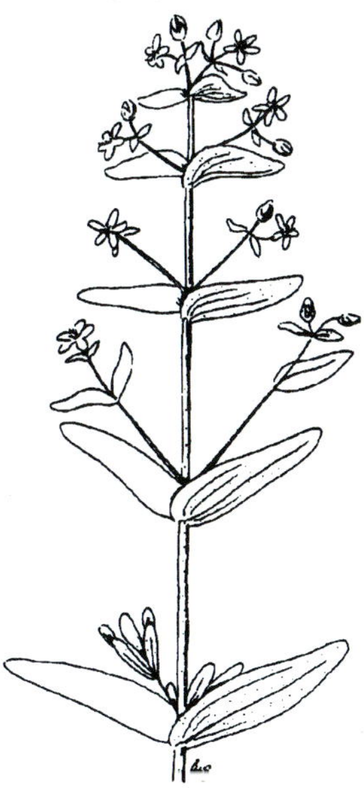

H. punctatum

H. perforatum

III. Necrosis and Sloughing: pigmentation pattern dependent; primary (direct) photosensitization; pruritis, erythema, swelling and sloughing of light-colored or thinly haired areas

Thamnosma
blisterweed
Rutaceae, citrus family

Signs and Diagnostic Aids

- severe pruritis, reddening, swelling, edema; sloughing of skin of teats, udder, escutcheon, less-pigmented and lightly haired areas
- severe cases: skin thickened, gelatinous appearance extending into deeper corium; crusty, ulcerative, or proliferative areas especially around the face
- lesion distribution coincides with lighter-colored areas
- may have diarrhea, restlessness, agitation, loss of appetite, and increased respiratory rate, occasionally violent behavior at disease onset
- sheep: lesions may be limited to ears, eyelids, face, lips, coronets
- signs typically last for 2 weeks, healing in 2 or more months
- eyes: photophobia, cloudy corneas, reddening of edges of cornea and conjunctiva

Pathology

- microscopically: necrosis, polymorphonuclear infiltrate in deeper corium

Treatment

- provide pain relief, soothing skin medications
- feed and water in shade
- nursing offspring may require supplemental feeding

Disease Problem

- mainly sheep; readily eaten
- contact with or consumption of fresh or dried plants
- 1% b.w. eaten for 1 to several days
- linear furanocoumarins activated in peripheral circulation by UV light to form transient reactive radicals that cause damage to surrounding tissues
- effects seldom lethal

Taxonomy and Ecology

- two species of toxicologic concern:

T. montana	turpentine broom
T. texana	blisterweed, Texas desert rue, ruda del monte, Dutchman's breeches

- herbs, subshrubs, or shrubs, 10–60 cm tall
- leaves simple, alternate, with blister-like glands
- flowers blue, purple, or yellow, petals 4
- fruits 2-lobed capsules
- dry, rocky slopes; southwestern quarter of continent

Thamnosma texana

T. texana

T. texana

T. texana

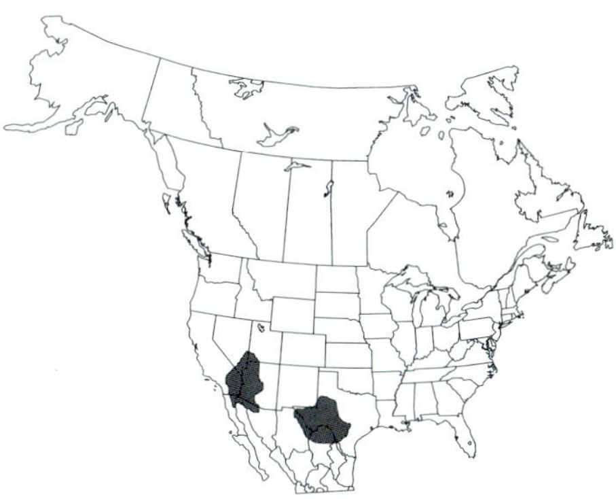

T. montana

IV. Necrosis and Sloughing: pigmentation pattern dependent; secondary (hepatogenous) photosensitization; pruritis, weakness, icterus, dark urine, weight loss

Agave
lechuguilla
Agavaceae, agave family

Signs and Diagnostic Aids

- mainly sheep and goats
- listless, not inclined to keep up with flock; stringy, thick mucoid discharge from eyes and nose, icterus of sclera and visible mucous membranes
- head and ears swollen, edema of face, lips, and underside of jaw; pruritis, purplish discoloration of coronary bands
- progressive debilitation, anorexia, weakness, emaciation, urine clear, dark yellow, or brown
- death in several days to a week or more
- increase in serum bilirubin and hepatic enzymes

Pathology

- thickening of skin and ears, cracking, sloughing of patches of skin; perhaps ear loss
- kidneys: swollen, greenish-black, numerous grey foci; liver brownish-yellow
- microscopically: renal tubules distended with albumin and casts of pigment and cellular debris; liver: fatty change, zonal necrosis, bile pigment in macrophages
- crystals or clefts in bile ducts; calcium salts of a sapogenin, probably smilagenin; also in bile canaliculi and Kupffer cells

Treatment

- provide supportive care, protect from sunlight

Disease Problems

- "swellhead," "goat fever," "lechuguilla fever"
- mainly sheep and goats; cattle less risk
- usually late winter or spring
- animals need not be eating *Agave* at time signs develop
- hepatotoxic saponin, smilagenin, renders liver incapable of eliminating phylloerythrin, which is activated by UV light to produce reactive radicals causing necrosis
- also affects urinary system

Taxonomy and Ecology

- one species of toxicologic concern:

  A. lecheguilla lechuguilla

- succulent perennial herb, 40–60 cm tall
- leaves long, spine-tipped; form basal rosette
- flowers numerous, yellow, borne on an elongated stalk 2–3 m tall
- open rocky sites, deserts; southwest Texas, New Mexico, and adjacent states in Mexico

Agave lechequilla

A. lecheguilla

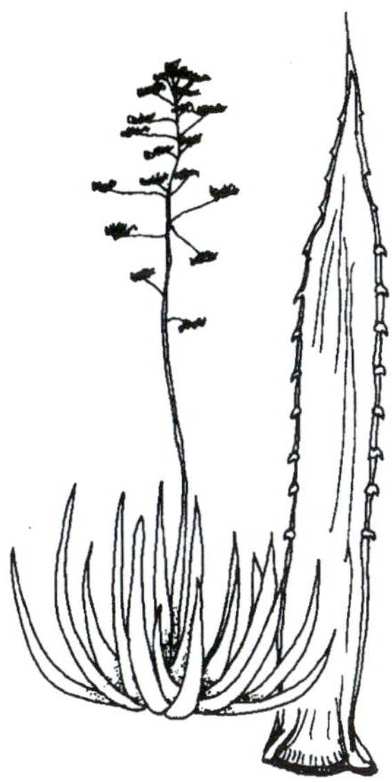

A. lecheguilla

IV. Necrosis and Sloughing: pigmentation pattern dependent; secondary (hepatogenous) photosensitization; pruritis, weakness, icterus, dark urine, weight loss

Nolina
sacahuista
Agavaceae, agave family

Signs and Diagnostic Aids
- mainly sheep and goats
- icterus; intensely yellow sclera and mucous membranes
- head and ears swollen; edema of the face, lips, and underside of the jaw; pruritis; purplish discoloration of coronary bands
- progressive debilitation, weakness, emaciation; urine clear, dark yellow, or brown
- often fatal in 1–2 weeks
- increase in serum calcium, bilirubin, and hepatic enzymes

Pathology
- thickening of skin and ears, cracking, sloughing of patches of skin; perhaps ear loss
- kidneys: swollen, greenish-black, numerous grey foci; liver brownish-yellow
- microscopically: renal tubules distended with albumin and casts of pigment and cellular debris; liver: fatty change, zonal necrosis, bile pigment in macrophages
- crystals or clefts in bile ducts; calcium salts of a sapogenin, probably smilagenin, also in bile canaliculi and Kupffer cells

Treatment
- provide supportive care, protect from sunlight

Disease Problems
- "bighead"; cattle seldom affected
- ingestion of flower buds, open flowers, and ripe fruit
- plants produce large numbers of flowers only every few years; useful forage in other years
- 1% b.w. eaten for 1 week or more
- toxicant unknown, likely hepatotoxic saponin; renders liver incapable of eliminating phylloerythrin, which is activated by UV light to produce necrosis
- also affects urinary system

Taxonomy and Ecology
- three species of toxicologic concern

N. bigelovii	Bigelow's nolina
N. microcarpa	sacahuista, small-seed nolina
N. texana	sacahuista, bunch-grass

- perennial subshrubs or herbs, to 1 m tall
- leaves simple, long, narrowly linear, spreading or arching, some spine tipped
- flowers in panicles, small, white, funnelform, on stem up to 1.8 m tall
- fruits ovoid, 3-lobed capsules
- Sonoran and Chihuahuan deserts of southwest

Nolina bigelovii

N. bigelovii

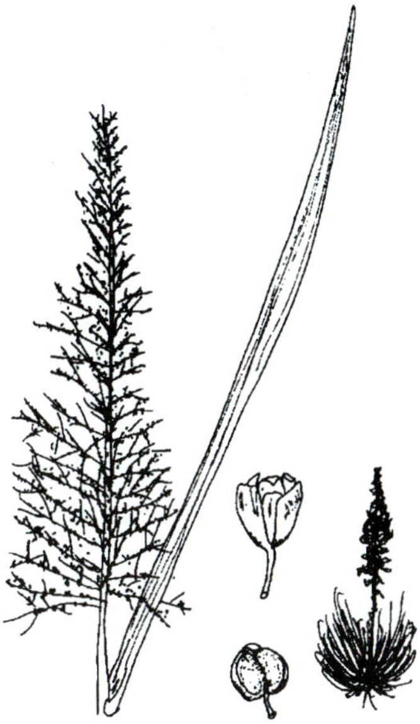

N. bigelovii

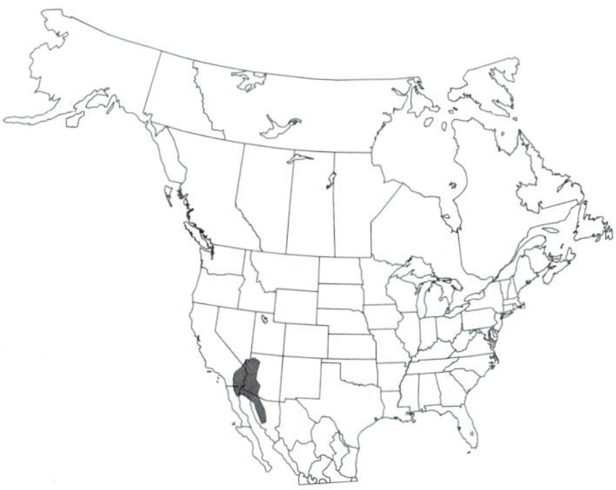

N. bigelovii

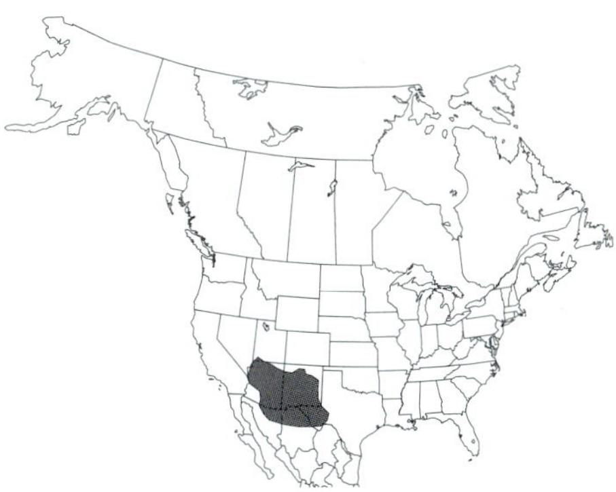

N. microcarpa

IV. Necrosis and Sloughing: pigmentation pattern dependent; secondary (hepatogenous) photosensitization; pruritis, weakness, icterus, dark urine, poor appetite, weight loss; sometimes neurologic signs in horses

Panicum
panicum
Poaceae, grass family

Signs and Diagnostic Aids

- onset abrupt over 1–2 days; rarely week or more
- sheep: swelling of head, loss of wool; sloughing of skin on ear tips, eyelids, around eyes, near lips, nose, and below the jaws; serum oozes forming scabby crusts; reddening of coronary bands, lameness, mild icterus
- horses: chronic poor appetite, weight loss; typically without photosensitization and liver involvement and only rarely hepatoencephalopathy, head pressing, periodic violent behavior
- increase in serum GGT, blood ammonia, total and direct bilirubin; severalfold decrease in BSP clearance

Pathology

- microscopically: mild necrosis of proximal convoluted tubules of kidneys; swelling, necrosis of hepatocytes, inflammation of bile ducts
- numerous birefringent needlelike crystals or clefts in small bile ducts, bile canaliculi, and Kupffer cells

Treatment and Prevention

- remove from forage; most animals recover
- provide nursing care, provide shade
- avoid grazing lambs or kids on *Panicum*
- graze mainly with cattle

Disease Problem

- "swellhead"; mainly lambs and kids
- less problem with mature sheep, Angora goats; cattle unaffected
- after grazing mature pasture for week or more
- horses: after grazing Kleingrass for up to 1–3 years
- crystalloid cholangiohepatopathy: calcium salts of glucuronidated saponin precipitated in bile ducts render liver unable to clear phylloerythrin, which is activated in peripheral circulation causing necrosis of tissues
- also affects respiratory and urinary systems

Taxonomy and Ecology

- representative species of toxicologic concern:

P. coloratum	Kleingrass
P. dichotomiflorum	fall panicum, smooth witchgrass, fall panic grass
P. maximum	guinea grass
P. miliaceum	common millet, broom millet, hog millet, proso millet

- annuals or perennials, 0.5–3.0 m tall
- inflorescences open-panicles
- spikelets dorsally compressed, florets 2, glumes unequal
- cultivated; mainly of southern half of continent

Panicum coloratum

P. coloratum

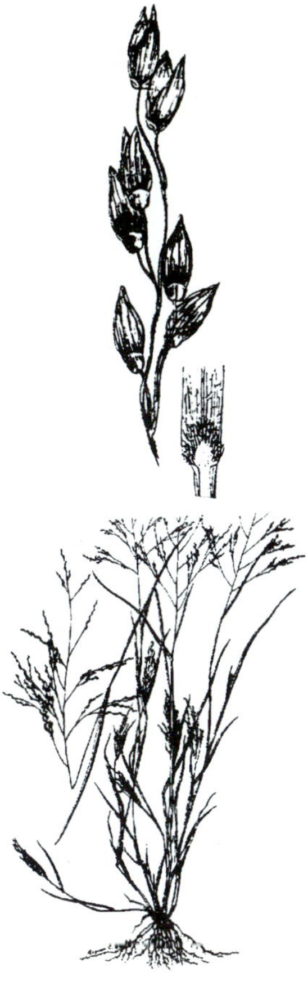

P. dichotomiflorum

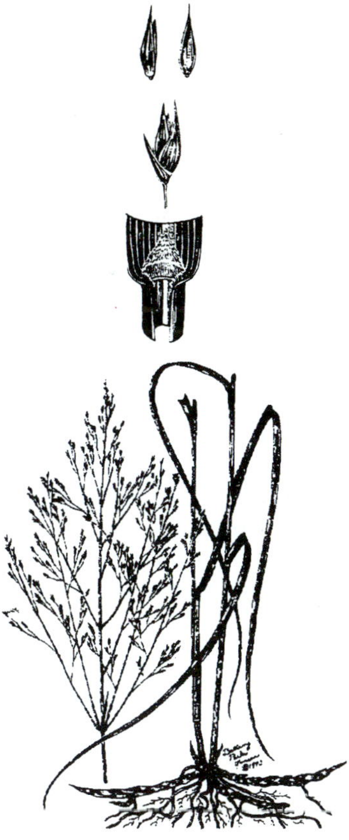

Panicum sp.

IV. Necrosis and Sloughing: pigmentation pattern dependent; secondary (hepatogenous) photosensitization; pruritis, weakness, icterus, dark urine, weight loss

Tribulus
goathead
Zygophyllaceae, caltrop family

Signs and Diagnostic Aids

- mainly sheep: icterus of sclera, visible mucous membranes
- head and ears swollen; edema of face, lips, and underside of jaw; pruritis, purplish discoloration of coronary bands
- progressive debilitation, weakness, emaciation, urine clear, dark yellow, or brown
- increase in serum bilirubin and hepatic enzymes

Pathology

- thickening of skin and ears, cracking, sloughing of patches of skin; perhaps ear loss
- kidneys: swollen, mottled brown
- liver: brownish-yellow, thickened bile duct walls
- microscopic: liver; moderate portal fibrosis and bile duct proliferation; polarizing crystals or clefts in and around bile ducts

Treatment

- provide supportive care, protect from sunlight

Disease Problems

- rare disease, most likely sheep
- all plant parts toxic
- crystalloid cholangiohepatopathy: calcium salts of glucuronidated saponin precipitated in bile ducts render liver unable to clear phylloerythrin, which is activated by UV light to produce reactive radicals in peripheral circulation causing necrosis of adjacent tissues

Taxonomy and Ecology

- two introduced species of toxicologic concern:

T. cistoides	Jamaican feverplant, bur nut
T. terrestris	goathead, puncture vine

- annual or perennial herbs, diffusely branched, stems prostrate to ascending to 1 m long
- leaves simple, opposite, 1 of each pair smaller or absent
- flowers solitary, sepals 5, petals 5, yellow, orange, or white
- fruits 5-angled schizocarps with 4–8 stout spines/horns
- disturbed soils; southern half of continent

Tribulus terrestris

T. terrestris

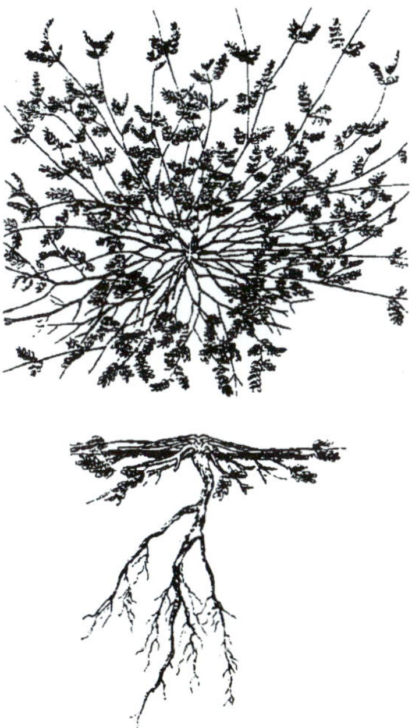

T. terrestris

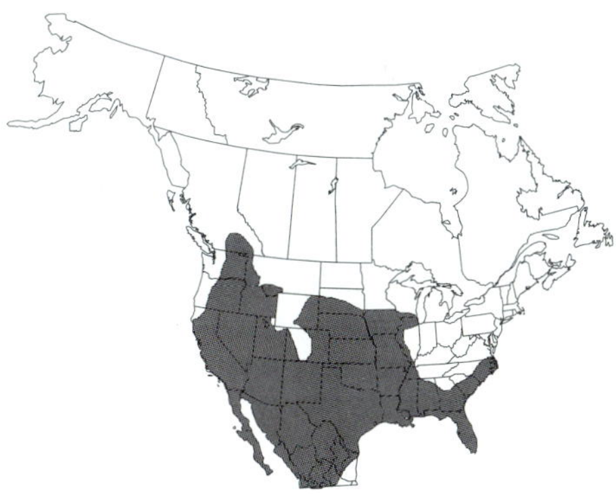

T. terrestris

V. Necrosis and Sloughing: without regard to pigmentation

Vicia
hairy vetch
Fabaceae, pea family

Signs and Diagnostic Aids
- rapidly progressive dermatitis; independent of pigmentation pattern
- papular swellings, plaques, thickening of the skin, exudative dermatitis, matting of hair by yellowish fluid, crusts covering areas of epithelial sloughing, pruritis, loss of hair
- skin thickened, fissured, or folded; lichenified
- initially changes on neck, tail, head, escutcheon, udder, and teats; progress to face, trunk, and limbs
- keratoconjunctivitis, and/or diarrhea ± blood, fever, cough, rapid breathing
- appetite reduced, rapid weight loss
- transient period of apparent recovery with renewed interest in eating after several days, but usual outcome death in 10–14 days

Pathology
- yellowish-gray infiltrate in dermis, adrenals, myocardium, renal cortex, which extends from surface of kidney; less commonly infiltrate in liver, spleen, lymph nodes, thyroid
- microscopically: infiltrate composed of macrophages, lymphocytes, plasma cells, eosinophils, multinucleated giant cells

Treatment
- provide nursing care; prognosis poor
- at first sign of disease, remove all animals immediately from pasture

Disease Problem
- consumption during periods of vigorous plant growth of various species
- occurs mainly in Oklahoma and Texas with *V. villosa*
- sporadic occurrence, often in pastures where *Vicia* grazed for many years without incident
- most cases in mature black cattle, Angus or Holstein; horses may also be affected
- signs appear 3–4 weeks or more after pasturing
- morbidity usually low (10%), but may be 40–50%; case fatality rate typically exceeds 50%
- toxicant and causative factors unknown

Taxonomy and Ecology
- one introduced species of toxicologic concern:

V. villosa hairy vetch

- herbaceous vine
- leaves 1-pinnately compound, terminal tendrils
- flowers pealike, blue and white to violet or rose
- cultivated; naturalized in waste areas, roadsides, fencerows; across continent

Vicia villosa

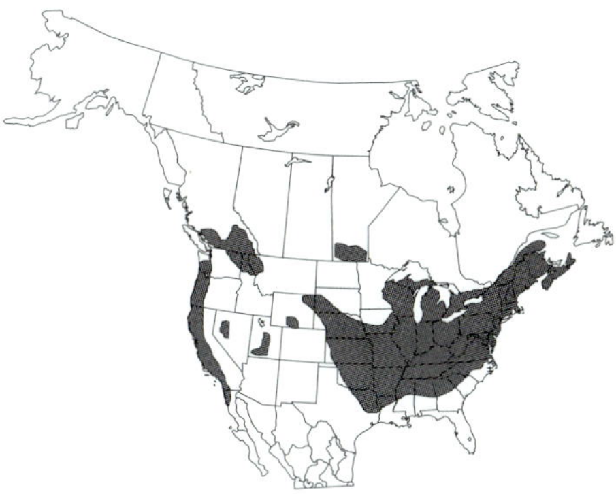

V. villosa

V. villosa

10
Plants Affecting the Urinary System

Thirty-one genera in 20 families affect the kidneys to varying degrees. Most are classified as of secondary importance because their toxicants directly affect other organs and systems, and the changes in the kidneys reflect the alterations of these other organs. Six genera in four families, however, are classified as primary because they affect the urinary system specifically and may pose a significant threat to livestock or pets.

Principal Signs

I. **Urine Consistency Thick:** abundant inflammatory cells, with cystitis and incontinence
II. **Urine Color Dark:** methemoglobin, myoglobin, or bile in urine
III. **Urine Color Red:** hemoglobin or blood in urine
IV. **Urine Odor Unusual**
V. **Urination Frequency Increased**
VI. **Urine Volume Increased:** renal tubular necrosis without oxalate crystals in urine
VII. **Urine Volume Increased:** renal tubular necrosis with oxalate crystals in urine

Plants Causing Problems

I. **Urine Consistency Thick:** abundant inflammatory cells, with cystitis and incontinence

Secondary

Sorghum	sorghum (page 18, 190)

II. **Urine Color Dark:** methemoglobin, myoglobin, or bile in urine

Secondary

Acer	maple (page 10)
Agave	lechuguilla (page 256)
Gutierrezia	broomweed (page 222)
Nolina	sacahuista (page 258)

Panicum	panicum (page 260)
Senna	senna (page 60)
Sorghum	sorghum (page 18, 190)
Thermapsis	goldenbanner (page 152)
Tribulus	goathead (page 262)

III. Urine Color Red: hemoglobin or blood in urine

Secondary	
Cynoglossum	hound's tongue (page 108)
Pteridium	bracken fern (page 8, 138)
Ranunculus	buttercup (page 56)
Senecio	groundsel (page 110)
Tertiary	
Leucaena	lead tree

IV. Urine Odor Unusual

Secondary	
Allium	onion (page 12)
Eupatorium	white snakeroot (page 206)
Isocoma	goldenbush (page 208)

V. Urination Frequency Increased

Secondary	
Conium	poison hemlock (page 204)
Eupatorium	white snakeroot (page 206)
Isocoma	goldenbush (page 208)
Nicotiana	tobacco (page 210)
Senecio	groundsel (page 110)
Trifolium	clover (page 76)
Tertiary	
Peganum	African rue

VI. Urine Volume Increased: renal tubular necrosis without oxalate crystals in urine

Primary	
Amaranthus	pigweed
Lilium	lily
Quercus	oak
Tertiary	
Anagallis	pimpernel
Hemerocallis	day lily

VII. Urine Volume Increased: renal tubular necrosis with oxalate crystals in urine

Primary	
Chenopodium	goosefoot
Halogeton	halogeton
Sarcobatus	black greasewood
Tertiary	
Atriplex	saltbush
Bassia	bassia
Oxalis	wood sorrel
Rheum	rhubarb
Rumex	sorrel
Salsola	Russian thistle

VI. Urine Volume Increased: renal tubular necrosis without oxalate crystals in urine; weakness, trembling, ataxia, perirenal edema

Amaranthus
pigweed
Amaranthaceae, amaranth family

Signs and Diagnostic Aids
- onset over hours to 1–2 days, weakness, trembling, unsteadiness when standing, ataxia, recumbency, almost flaccid paralysis
- coma, death in 1–2 days
- case fatality rate in excess of 50%
- increase in BUN to 90–200+ mg/dl; increase in potassium

Pathology
- pale yellow edema fluid around kidneys and omentum
- kidneys pale, normal in size or swollen
- free fluid in abdominal cavity and other areas
- microscopically: renal interstitial edema, tubular epithelial degeneration and necrosis, proximal tubules dilated, filled with casts or proteinaceous material

Treatment
- provide nursing care to allow recovery of tubular function

Disease Problem
- "perirenal edema"
- pigs, cattle, sheep
- large quantities of fresh plants eaten for 5–10 days, summer and autumn, most commonly caused by *A. retroflexus*
- excess serum potassium leads to cardiac failure
- toxin unknown, but not oxalate salts
- also affects blood and neuromuscular system

Taxonomy and Ecology
- representative species of toxicologic concern:

A. hybridus	smooth pigweed, green pigweed, carelessweed, quelite morado
A. palmeri	Palmer pigweed, Palmer amaranth, carelessweed
A. powellii	Powell pigweed, Powell amaranth
A. retroflexus	redroot pigweed, rough pigweed, quelite
A. spinosus	spiny pigweed, quelite espinoso

- annual herbs, to 3 m tall
- leaves simple, alternate, some with axillary spines
- flowers staminate and pistillate, in terminal or axillary spikes
- disturbed sites and waste areas across continent

Amaranthus retroflexus

A. retroflexus

A. retroflexus

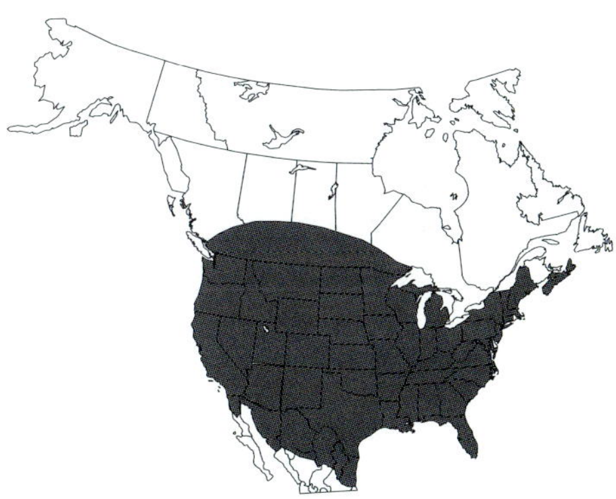

A. retroflexus

VI. Urine Volume Increased: renal tubular necrosis without oxalate crystals in urine; cats; vomiting, depression, weakness, trembling, ataxia

Lilium
lily
Liliaceae, lily family

Signs and Diagnostic Aids
- cats only
- onset within 2–4 days after ingestion
- vomiting, depression, loss of appetite
- death within 5 days if not treated vigorously
- marked increase in blood creatinine, BUN, phosphorus, potassium

Pathology
- yellowish fluid throughout body
- kidneys slightly swollen
- microscopically: dilation and necrosis of renal proximal tubules, flattened epithelium

Treatment
- administer emetics, activated charcoal, saline cathartic
- provide fluid diuresis immediately
- imperative to treat vigorously in first 24 hours to ensure survival

Disease Problem
- all plant parts toxic
- single flower may cause intoxication
- at present only noted in cats, perhaps only a matter of dosage
- toxin unknown, but not oxalate crystals
- all lilies, including *Hamerocallis*, day lily, should be considered a risk

Taxonomy and Ecology
- introduced ornamental species of toxicologic concern:

L. candidum	Madonna-lily, white-lily
L. lancifolium	tiger-lily, devil-lily
L. longiflorum	Easter-lily, trumpet lily
L. pardalinum	leopard-lily, panther-lily

- erect herbs, 50–250 cm tall, from fleshy white, yellow, or purple bulbs
- leaves long, basal
- flowers showy, typically funnelform, of various colors
- cultivated across continent; seasonal houseplants

Lilium longiflorum

L. longiflorum

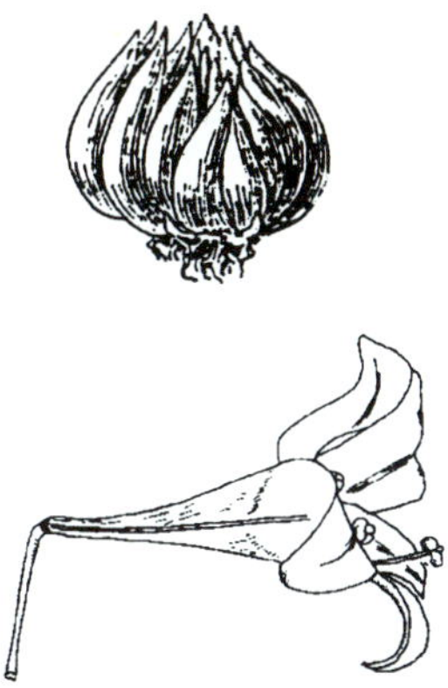

L. longiflorum

L. pardilinum

VI. Urine Volume Increased: renal tubular necrosis without oxalate crystals in urine; weakness, trembling, ataxia, rumen stasis, diarrhea; subcutaneous edema of neck, brisket, perineum

Quercus
oak
Fagaceae, beech family

Signs and Diagnostic Aids
- initially—anorexia, listlessness, rumen stasis, constipation
- later—diarrhea (± bloody), dehydration, colic; subcutaneous edema of the neck, brisket, abdomen, perineum; increased frequency of urination
- increase in BUN, serum creatinine, potassium, phosphate; decrease in sodium, chloride, and calcium
- urine analysis: low specific gravity (1.010), proteinuria, glucosuria, hematuria
- sheep and goats: edema minimal
- horses: diarrhea (± blood), colic, consistent and severe tenesmus

Pathology
- cattle: ascites, hydrothorax, hydroperitoneum, subcutaneous edema; edema of mesentery (especially perirenal) and gut mucosae
- kidneys slightly swollen and pale, with petechiae on surface extending into cortex
- microscopically: diffuse renal tubular dilation, necrosis of cortical tubular epithelial cells
- horses: gut lesions more severe, edema of caecal, colonic mucosa; renal disease less severe

Treatment and Prevention
- administer fluids, electrolytes
- supplement with alfalfa or feed containing 5–10% calcium hydroxide prior to exposure to *Quercus*
- administer activated charcoal, polyvinylpyrolidone or 4,000 MW polyethylene glycol (PEG) orally
- reduce grazing pressure, allow more forage selectivity

Disease Problem
- cattle; less commonly in sheep, goats, horses
- develops when animals forced to eat half or more of diet of young leaves, buds, and/or acorns for several days or more; 6% b.w. over several days
- mature leaves are of much reduced risk
- hydrolyzable tannins: hydrolyzed by gut esterases to polyphenolics; gallic acid, digallic acid, ellagic acid
- astringent effects on gut mucosa, tissue destruction at sites of high acid concentrations, liver or kidney tubule

Taxonomy and Ecology
- representative species of toxicologic concern:

Q. douglasii	blue oak, mountain white oak, iron oak
Q. gambelii	Gambels oak, Utah white oak, Rocky Mountain white oak
Q. garryana	Oregon white oak, garry oak, Oregon oak
Q. havardii	shinnery oak, shin oak
Q. marilandica	blackjack oak, jack oak, blackjack
Q. turbinella	shrub live oak, Sonoran scrub oak

- shrubs or trees, 1–40 m tall
- leaves diverse shape, some with spine-tipped lobes
- staminate flowers in long catkins as leaves emerge
- fruits nuts with scaly caps (acorns)
- numerous species across continent; most are toxic

Quercus

Q. marilandica

Q. havardii

Q. gambelii

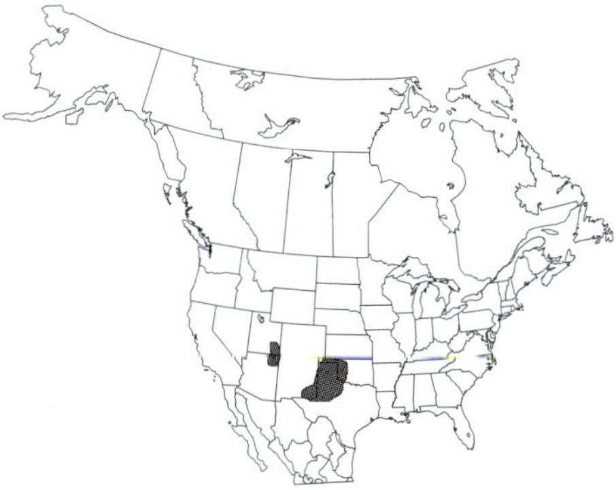

Q. havardii

VII. Urine Volume Increased: renal tubular necrosis with oxalate crystals in urine; abrupt onset; depression, weakness, tremors, ataxia, labored respiration

Chenopodium
goosefoot
Chenopodiaceae, goosefoot family

Signs and Diagnostic Aids

- abrupt onset; depression, weakness, tremors, ataxia, weak pulse, and labored respiration; hypocalcemia-type signs
- prostration, coma, death in a few hours
- marked decrease in serum calcium

Pathology

- kidneys: pale, swollen
- excess fluid in body cavities, edema of various tissues
- microscopically: nephrosis with tubular epithelial desquamation, tubule dilation, atrophy

Treatment

- administer parenteral calcium solutions; relapses may occur

Disease Problem

- rare; mainly sheep
- oxalate accumulation in plant
- risk mainly when large amounts eaten quickly; moderate amounts eaten repeatedly little problem
- calcium oxalate precipitation in blood and tissues
- also affects blood (nitrate accumulation)

Taxonomy and Ecology

- representative species of toxicologic concern:

C. album	lambsquarters, white goosefoot, quelite cenizo, nabo blanco
C. ambrosioides	wormseed, Mexican tea, Spanish tea
C. carinatum	keeled goosefoot, ridged goosefoot
C.glaucum	oak-leavedgoosefoot

- annual or perennial herbs, erect, prostrate, or decumbent, 10–300 cm tall, often turning red- or maroon-striped with maturity
- leaves simple, alternate, primarily cauline
- flowers in numerous glomerules, in terminal or axillary spikes or panicles
- disturbed sites and waste areas across continent

Chenopodium album

C. album

C. album

VII. Urine Volume Increased: renal tubular necrosis with oxalate crystals in urine; abrupt onset; weakness, coughing, ataxia, rapid shallow respiration, extensor rigidity

Halogeton
halogeton
Chenopodiaceae, goosefoot family

Signs and Diagnostic Aids
- sheep: dullness, lowering of head, loss of appetite, coughing, drooling with white froth around the mouth
- later: increase in severity; weakness, stiffness, ataxia of pelvic limbs, rapid/shallow breathing
- eventually: coma, with jerky extensor rigidity, irregular respiration; death may occur within a few hours
- little tendency for tremors to occur
- cattle: subacute locomotor difficulties mainly when forced to move; stiffness first in pelvic limbs, then all limbs; recumbent but recover if given feed, water
- decrease in blood calcium, to 1.4 mg/dl, magnesium; twofold increase in phosphorus; moderate increase in BUN

Pathology
- free fluid: abdominal, chest cavities, heart sac
- kidneys: pale, swollen/edematous, birefringent crystals from scrapings of cut surfaces
- microscopically: renal tubules filled with proteinaceous casts and calcium oxalate crystals; epithelium flattened and necrotic; edema and hemorrhage ruminal wall; oxalate crystals in submucosa and walls of vessels

Treatment and Prevention
- parenteral calcium (calcium borogluconate i.v.) therapeutic for cattle; delay of deaths of sheep but not survival
- preventive calcium supplements: pellet of 83% alfalfa, 15% calcium carbonate, and 2% molasses prior to exposure
- calcium chloride or dicalcium phosphate: protection short-lived, effective shortly before oxalate challenge

Disease Problem
- mainly sheep, less severe in cattle
- no problem if eaten in moderation with other forage
- abrupt onset when large amounts eaten in short period with rumen empty
- sodium oxalate salts; more severe neurologic effects
- precipitation with calcium in blood and tissues

Taxonomy and Ecology
- one introduced species of toxicologic concern:

  H. glomeratus halogeton

- annual herb from taproot, 10–50 cm tall, dull green to blue-green herbage when young; stems profusely branched at bases
- leaves alternate, fleshy, bristle-tipped, tiny, sausage-like
- flowers inconspicuous in axillary glomerules
- desert saltbush and sagebrush communities of western part of continent

Halogeton glomeratus

H. glomeratus

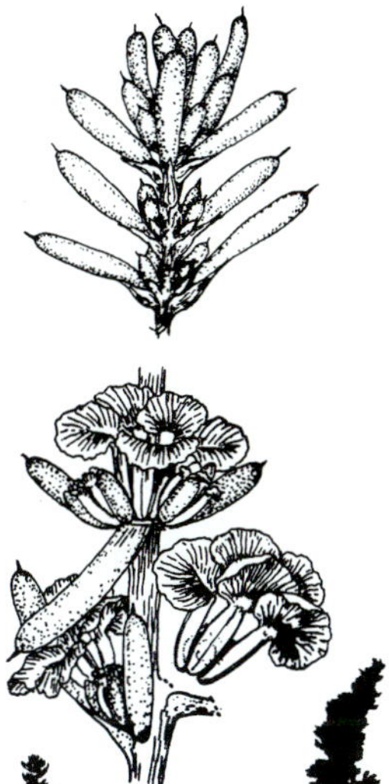

H. glomeratus

H. glomeratus

VII. Urine Volume Increased: renal tubular necrosis with oxalate crystals in urine; abrupt onset; depression, weakness, labored respiration

Sarcobatus
black greasewood
Chenopodiaceae, goosefoot family

Signs and Diagnostic Aids
- abrupt onset; depression, weakness, weak pulse, labored respiration, tremors
- prostration, coma, and death in a few hours
- marked decrease in serum calcium

Pathology
- kidneys: pale and swollen
- excess fluid in body cavities
- microscopically: renal tubular dilation and flattened degenerated tubular epithelium with numerous oxalate crystals in tubules and in walls of small arteries in rumen and brain

Treatment
- parenteral calcium solutions: temporary relief of signs; animals may become ambulatory within a few hours
- relapses occur; death due to neurologic effects and renal failure

Disease Problem
- browsed extensively by sheep typically without problem
- losses when hungry/thirsty animals eat large amounts quickly
- mixture of sodium and potassium oxalates
- precipitation with calcium in blood and tissues
- also affects neuromuscular system (tremors)

Taxonomy and Ecology
- one species of toxicologic concern

S. vermiculatus	greasewood, black greasewood

- erect, deciduous, shrubs, 0.3–3 m tall
- branches stout, rigid, spine-tipped
- leaves opposite below and alternate above, succulent
- flowers imperfect; staminate in terminal spikes, pistillate in leaf axils
- highly alkaline soils and high water tables; semiarid valleys of intermountain west

Sarcobatus vermiculatus

S. vermiculatus

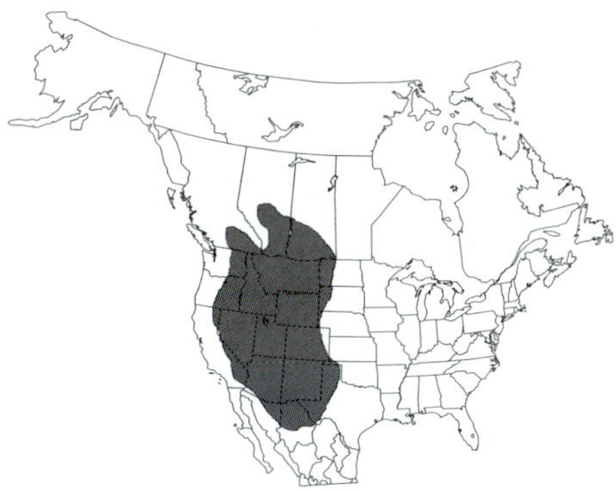

S. vermiculatus

S. vermiculatus

11
Plants Causing Sudden Death

Perhaps nothing is more disheartening to a rancher, farmer, or pet owner than the unexpected discovery of a dead animal that appeared healthy only a few hours earlier. The emotional loss of a living organism is coupled with economic loss and, if many animals are involved, the effects can be financially devastating. Fortunately, the majority of toxic plants causing death produce clinical signs over a period of time—from hours to days or weeks—that alert the producer or owner to the problem of intoxication, especially if the animals are being observed at periodic intervals.

The frequency of observation is the critical determinant in differentiating between toxic plants that indeed cause death quickly, and those that cause death slowly. Four groups of plants are most likely to cause death within a few hours of plant ingestion. They are the:

nitrate-accumulating plants	pages 4
cyanogenic plants	pages 4, 118
plants causing cardiac failure	pages 21–22
plants causing seizures	pages 117–118

As is apparent, we have not designated genera or species of these groups as primary, secondary, and tertiary causative plants as was done for the taxa in the preceding chapters. Each taxon of these groups is capable of causing sudden death, regardless of the frequency that they do so. However, with the exception of *Taxus*, in plants causing cardiac failure, signs suggestive of intoxication are typically apparent for at least a few minutes and in many cases for a few hours.

Again, we must emphasize that the occurrence of "sudden death" due to toxic plants is a reflection of the livestock or pet owner's management practices and the frequency of observing animals.

Glossary: Botanical Terms

achene—small, dry, indehiscent, 1-seeded fruit with ovary wall free from the seed except at funiculus.

alternate—(1) arrangement of organs such as leaves, branches, flowers, or fruits, one per node; (2) arrangement of organs in one floral whorl between organs of adjacent floral whorl; e.g., petals alternate with sepals.

apex (pl. *apices*)—morphological tip of an organ.

aril—fleshy appendage partially or entirely enveloping seed.

awn—bristlelike appendage extending beyond body of organ, usually at tip or from dorsal surface.

axil—upper angle formed between axis and any organ arising from it; e.g., between leaf and stem.

axillary—borne in axil of leaf.

basal—attached or situated at base of plant or organ.

beak—long, pointed, terminal projection of fruit or seed.

berry—fleshy, indehiscent fruit derived from one pistil and having a pulpy or fleshy pericarp; e.g., grape or tomato.

bilabiate—two-lipped.

bilateral—divisible into mirror-image halves in only two planes; cf. radial.

bract—modified leaf subtending flower or an inflorescence.

bristle—stiff hair.

bulb—short, thick, subterranean stem bearing fleshy leaves; e.g., onion.

capsule—dry, dehiscent fruit derived from a compound pistil and splitting along two or more lines of suture.

cauline—pertaining to arising from aerial stem.

ciliate—with marginal fringe of hairs.

claw—long, narrow, petiole-like base of petals or sepals in some flowers.

colonial—forming colonies, typically asexually via rhizomes, stolons, or root sprouts.

compound—organ divided into two or more similar parts; e.g., compound leaves or compound pistils.

compressed—(1) flattened in one plane; (2) in Poaceae (grasses), two types recognized: laterally—spikelets flattened along margins (sides) of glumes and lemmas; and dorsally—spikelets flattened along midnerves (backs) of glumes and lemmas.

corona—crown or cuplike appendage or ring of appendages arising from petals or stamens in *Narcissus* (daffodil) or *Asclepias* (milkweed).

crown—(1) base of tufted, herbaceous perennial where shoot and root systems merge; (2) overall appearance or form of branches of tree or shrub.
cyme—broad, more or less flat-topped determinate inflorescence, with central or terminal flowers opening first.
deciduous—(1) not persistent, falling off when no longer functional; e.g., leaves falling in autumn or floral parts after anthesis; (2) pertaining to a plant that drops all of its leaves annually.
decumbent—lying horizontally on ground but with terminal portion ascending.
dentate—pertaining to a margin with sharp teeth pointing outward rather than forward.
dioecious—with pistillate and staminate flowers borne on different plants; cf. monoecious.
disk floret—in Asteraceae (sunflowers), perfect floret with tubular corolla; cf. ray floret.
dissected—deeply divided or cut into many segments.
dorsal—pertaining to or located on back; in Poaceae (grasses), one type of compression, with spikelets flattened along midnerves (backs) of glumes and lemmas; cf. lateral.
drupe—fleshy, 1-seeded, indehiscent fruit with seed enclosed in hard or papery endocarp called the pit or stone; e.g., *Prunus persica* (peach).
endocarp—inner layer of fruit wall or pericarp that is differentiated into three layers.
entire—pertaining to a continuous margin, without teeth or divisions.
escaped species—cultivated taxon establishing plants or populations in natural habitats.
evergreen—(1) plant having foliage that remains green for more than one growing season and not falling en masse; (2) never lacking green leaves; (3) remaining green through winter.
fascicle—tight cluster or bundle of organs, appearing to arise from a common point.
floret—(1) in Poaceae (grasses), flower and two associated bracts, the lemma and palea; subunit of grass spikelet; (2) in Asteraceae (sunflowers), one flower of a composite head with five fused petals, a highly modified calyx, five stamens fused by their anthers, and an inferior ovary; three types recognized—disk, ray, and ligulate.
foliage—collective term for leaves of plant.
follicle—dry, dehiscent fruit derived from simple pistil and dehiscing along one line of suture.
free—separate, not fused or attached to another organ.
frond—leaf of ferns.
fruit—mature ovary or ovaries, plus any other structures maturing with them after fertilization.
fused—parts or organs joined to one another.
fusiform—spindle-shaped; broadest near middle and tapering toward ends.
glabrous—devoid of pubescence or hairs of any form.
gland—(1) appendage, protuberance, or depression on surface of organ or at end of hair that secretes usually sticky fluid; (2) protuberance that may not be secretory; e.g., warty swelling at base of leaf blade in *Prunus* (plum).
globose—spherical or nearly so; globular.

glomerule—compact, cymose cluster of flowers or heads subtended by involucre.

glume—in Poaceae (grasses), one of a pair of bracts at base of spikelet, typically dry and membranous.

habit—general appearance of plant; may refer to single character or collectively to all characters.

habitat—environmental conditions or type of area normally occupied by a plant; e.g., borrow ditch, tallgrass prairie, or waste area.

head—(1) dense spherical, hemispheric, or flat-topped inflorescence of sessile or subsessile flowers; (2) in Asteraceae (sunflowers), dense spherical, hemispheric, or flat-topped inflorescence of sessile flowers borne on a common receptacle and subtended by involucre.

herb—plant with nonwoody aerial stems that die at end of growing season.

husk—covering, generally derived from perianth or involucre, that encloses fruit.

imperfect—pertaining to a flower that lacks either stamens or pistil or both; cf. perfect.

inflorescence—arrangement of flowers and accessory parts on an axis.

introduced species—taxon occurring in an area due to human activity; not indigenous; cf. native, naturalized.

lanceolate—lance-shaped; rather narrow, tapering to both ends with broadest part below middle.

lateral—(1) on or at side of axis or organ; (2) in Poaceae (grasses), one type of compression, with spikelets flattened.

leaflet—single segment of compound leaf.

legume—in Fabaceae (legumes), dry, dehiscent, 1-locular, 2-valved fruit derived from a simple pistil and opening along two sutures.

lemma—in Poaceae (grasses), outer and generally large bract of two subtending flowers; one of two bracts of a floret.

ligule—in Poaceae (grasses), membranous or hairy appendage on adaxial surface of leaf at junction of sheath and blade.

margin—edge or border of organ.

monoecious—with pistillate and staminate flowers borne on same plant; cf. dioecious.

native species—taxon occurring naturally in an area; presence not due to human activity; indigenous; cf. introduced, naturalized.

naturalized species—introduced taxon established in an area and growing and reproducing as if it were a native; cf. native.

nut—dry, indehiscent, usually 1-loculed, 1-seeded fruit derived from compound ovary with bony, woody, leathery, or papery pericarp and partially or wholly enclosed in involucre.

nutlet—(1) dry, indehiscent, 1-seeded fruit derived from one-half of carpel; characteristic of Lamiaceae (mints), Boraginaceae (borages), and Verbenaceae (vervains).

oblong—rectangular but with rounded ends; two to three times longer than broad and with sides nearly parallel.

opposite—(1) arrangement of organs such as leaves, branches, flowers, or fruits two per node; (2) arrangement of organs in one floral whorl superposed with those of adjacent floral whorl; e.g., stamens opposite the petals.

ovate—egg-shaped in two dimensions and attached at broader end; applied to plane surfaces.
palmate—with two or more lobes or segments radiating from the same basal point; resembling fingers spreading from palm of hand.
panicle—indeterminate, branched inflorescence comprising rachis, one to several series of branches, and pedicels bearing flowers.
pedicel—(1) stalk bearing an individual flower, or the ultimate axis of compound inflorescence; (2) in Poaceae (grasses), stalk bearing a spikelet.
perfect—pertaining to flowers with both stamens and pistils present and functional; cf. imperfect.
petaloid—petal-like in color and/or shape and/or texture.
petiole—stalk of leaf connecting blade and stem node.
pinnate—feather-like; with closely arranged secondary veins radiating from midvein (primary) in two rows.
pinnately—descriptive of a compound leaf with leaflets borne in two rows along an axis; if 1-compound, leaflets are attached to rachis; if 2- or 3-compound, leaflets are attached to secondary or tertiary axes (rachillas).
pistil—innermost floral organ; normally differentiated into ovary, style, and stigma; female organ; when simple, comprising one carpel; when compound, comprising two or more carpels.
pistillate—pertaining to flower that bears pistil or pistils but lacks functional stamens; cf. staminate.
prostrate—lying flat on ground.
raceme—indeterminate inflorescence consisting of a rachis bearing a number of pedicelled flowers.
radial—(1) denoting radius; (2) developing uniformly on all sides; (3) divisible into mirror-image halves in multiple planes.
rank—a vertical row along an axis.
ray floret—in Asteraceae (sunflowers), pistillate or neutral floret with strap-shaped corolla; cf. disk floret.
rhizome—horizontal underground stem producing aerial stems and leaves along its length or end, with scale leaves present or absent.
rosette—cluster of leaves radiating from common center or crown; usually at or near ground level.
rotate—pertaining to circular corolla shape with short tube and wide limb(s) spreading at right angles to tube.
samara—winged, dry, indehiscent, 1-seeded, achenelike fruit; e.g., *Ulmus* (elm) and *Fraxinus* (ash).
schizocarp—dry, dehiscent fruit derived from a compound pistil that splits into separate 1-seeded, 1-carpellate mericarps at maturity.
serrate—with sharp teeth pointing forward on margin of organ.
shrub—woody plant with multiple stems from base but lacking a single trunk.
simple—pertaining to an organ not divided into distinct parts or segments; not compound; e.g., leaf with undivided blade or pistil with one carpel; cf. compound.
spicate—spikelike; inflorescence with appearance of spike but flowers not truly sessile.
spike—inflorescence consisting of central rachis bearing one or more sessile flowers.

spikelet—in Poaceae (grasses), highly modified inflorescence typically consisting of two glumes, one or more florets, and rachilla.

staminate—pertaining to a flower that bears stamens but lacks a functional pistil; cf. pistillate.

stipule—one of pair of appendages borne at base of leaf.

stone—hard, woody endocarp of drupe; a pit.

symmetry—condition whereby half of organ or structure is mirror image of other half; correspondence in size, shape, and relative position of parts of two halves.

taproot—persistent, well-developed primary root, generally larger than secondary roots that arise from it.

taxon (pl. taxa)—general taxonomic term for any group of any rank.

tendril—rotating or twisting threadlike process or extension by which a plant grasps an object for support; may be a modified stem, leaf, leaflet, or stipule.

terminal—tip or apex of stem, summit of axis; e.g., terminal leaflet of 1-pinnately compound leaf.

toothed—general term encompassing margin with serrations, dentations, or crenations.

tuft—dense cluster of organs or structures.

umbel—indeterminate, flat-topped or hemispheric inflorescence consisting of flowers borne on pedicels arising from common point of attachment; in simple umbels, pedicels arise from end of peduncle; in compound umbels, pedicels arise from end of rays that arise from end of peduncle.

umbellet—one of ultimate umbellate clusters of compound umbel.

vein—vascular bundle visible externally; primary veins are largest, occurring singly as midvein or series; secondary veins are next size class, arising from primary veins.

venation—arrangement of veins on surface of organ.

viscid—covered with sticky or gelatinous exudate; gummy to the touch.

weed—plant species that aggressively colonizes disturbed habitats and cultivated lands.

whorl—(1) arrangement of organs such as leaves, branches, flowers, or fruits three or more per node; (2) two or more cyclic groups of sepals or petals or stamens.

wing—thin, flat, membranous expansion forming ridge or flange on organ.

Index